𝕱𝖆𝖑𝖑𝖎𝖓𝖌 𝕬𝖕𝖕𝖑𝖊 𝕻𝖗𝖊𝖘𝖘

IEP, Inc.

Palm Springs, CA

ISBN 9781508426974

Table of Contents

Real World Activities

Forward

For those who choose to use this Physics Laboratory Manual, please recognize it is a work in progress. This edition is a first attempt to provide laboratory experiments, which demonstrate and enhance the understanding of the concepts and principles taught in the AP Physics 1 course promulgated by the College Board.

Some experiments are qualitative in nature while many others are quantitative, or a mixture of both. The format and level of detail also varies from lab experiment to lab experiment. This is a deliberate attempt to challenge students with a variety of laboratory procedures, forcing them to adapt and improvise as an intentional part of the laboratory experience.

The laboratory experiments are also intentionally designed with a minimum of exotic equipment and materials required. Most of the equipment and materials can be obtained by visiting the local hardware or building supply store. This forces the students to assemble the experimental set-up, think through the procedures and possible sources of error, and refine their execution of the lab without pre-assembled equipment and sensor- driven data collection.

As the College Board provides AP Physics 1 guidance about course content and laboratory experiments to support the course, this manual will be updated. Also through the experience of doing each of these experiments with my students, I will be improving experimental procedures, clarity of instructions, and the format of the manual in each subsequent edition.

Any mistakes or conceptual errors are entirely my own. I covet your responses to the experiments contained herein and will incorporate suggestions into future editions.

01 July 2015

Gregory Hepner
San Diego, California

Welcome to the Physics Laboratory

Physics is our human attempt to explain the workings of the world. The success of that attempt is evident in the technology of our society. You have already developed your own physical theories to understand the world around you. Some of these ideas are consistent with accepted theories of physics while others may not be. This laboratory manual is designed, in part, to help you recognize where your ideas agree with those accepted by physics and where they do not. It is also designed to help you become a better physics problem solver.

You are presented with physical theories in the classroom and in your textbook. In the laboratory you will apply the theories to real-world problems by comparing your application of those theories with reality. You will clarify your ideas by: answering questions and solving problems before you come to the lab; performing experiments and having discussions with classmates in the lab; and writing lab reports after you leave. Each laboratory has a set of problems that ask you to make decisions about the real world. As you work through the problems in this laboratory manual, remember the goal is not to make lots of measurements. The goal is to examine your ideas about the real world.

The four components of the course - lecture, discussion, problem analysis, and laboratory – each serves a different purpose. The laboratory is where physics ideas, often expressed in mathematics, meet the real world. Because labs meet on different days of the week, you may deal with concepts in the lab before meeting them in the lecture. In that case, the lab will serve as an introduction to the lecture. In other cases the lecture will be an introduction to the lab.

The amount you learn in lab will depend on the time you spend in preparation before coming to the lab. Before the lab you must read the appropriate sections of your text, read the assigned lab manual experiment to develop a fairly clear idea of what will be happening, and consider the prediction/questions for the lab.

Your lab group may be asked to present predictions and data to other groups so everyone can participate in understanding how specific measurements illustrate general concepts of physics. You should always be prepared to explain your ideas or actions to others in the class. To show your instructor you made the appropriate connections between your measurements and basic physical concepts, you will write a laboratory report. Guidelines for preparing lab reports are found in the next section. An example of a good lab report is shown. Do not hesitate to discuss any difficulties with the instructor.

Relax. Explore. Make mistakes. Learn from them. Ask lots of questions. Have fun.

To Be Successful in This Laboratory

Safety is first in any laboratory!
If in doubt about any procedure, or if it seems unsafe to you, **STOP!** Ask your instructor for help.

What to Bring to Each Laboratory Session
- A graph-ruled laboratory journal to all lab sessions. Your journal is your "extended memory" and should contain everything you do in the lab and all your thoughts as you are going along. Your lab journal **must be bound** and must **not** allow pages to be easily removed (as with spiral bound notebooks).

- A "scientific" calculator.

- This laboratory manual.

Your Laboratory Journal
Keeping a neat and complete laboratory journal is an essential skill for this course. The ability to keep a good journal will help you in your future academic and professional career.

All of your original work must be preserved in your lab journal. **Never tear pages out of your lab journal.** When you make a mistake, neatly cross out that part. Make sure you can still read it, in case there is useful information there. When asked to turn in copies of work from your journal either make photocopies or turn in the original journal.

All your raw data, calculations, and conclusions must be recorded in your lab journal. You must use a **bound** **graph-ruled** journal for this course. The lab journal is where you record all activities related to the lab, including initial calculations and/or analysis.

It is useful to keep a few pages at the beginning of the journal blank to later fill them as a table of contents. For the purpose of organization, skip a page at the end of one lab and start the next lab with a title. You should include not only all raw data, graphs, etc...but also sketches of the experimental setup with appropriate explanations. Graphs must have properly labeled axis with units. You should include the numerical data in addition to the graphs. Computers/tablets fail. Do not depend on them to retain your data. Write important things!

Remember, it is difficult to anticipate what information will or will not be needed for later analysis. It is better to record too many details than not enough. The only thing entered in your lab journal before a particular lab should be any required warm-up questions and/or prediction. The lab journal should be a running record of what is done in the course of the laboratory experiment.

Prepare for Each Laboratory Session

Each laboratory consists of a series of related problems that can be solved using basic concepts and principles. Sometimes all lab groups will work on the same problem, other times groups will work on different problems and share results.

- Before beginning a lab, carefully read the Purpose/Problem and Procedure sections.

- Each lab may contains several different experimental problems. Before you come to a lab, complete the *Prediction* and *Method Questions,* if assigned. The method questions help build a prediction for the given problem. It is usually helpful to answer the method questions before making the prediction. **The predictions may be checked (graded) by the instructor at the beginning of each lab session.** This preparation is crucial if you are going to understand your laboratory work There are two other reasons for preparing:
  - There is nothing duller or more exasperating than plugging mindlessly into a procedure you do not understand.
  - The laboratory work is a **group** activity where every individual contributes to the thinking process and activities. Other members of your group will be unhappy if they must consistently carry the burden of one who is not doing his/her share.

Attendance

Attendance is required at all labs **without exception**. If something disastrous keeps you from your scheduled lab, contact the instructor **immediately**. Most labs require two or three individuals to execute the experiment, therefore *missed labs are extremely difficult to make up.*

Laboratory Reports

At the end of every lab experiment, write up the experimental results. Your lab report must be a clear and accurate account of what you and your group members did, the results you obtained, and what the results mean. The lab report must not be copied or fabricated. (That would be scientific fraud.) **Copied or fabricated lab reports will be treated as cheating on a test and will result in referral to the Dean of Students.** Your lab report should describe your predictions, experiences, observations, measurements, and conclusions. A detailed description of the lab report is discussed in the next section of this manual.

Grades

Satisfactory completion of the lab is required as part of your course grade. The laboratory grade is 25% of your final course grade. Once again, **each lab report is due, without fail, within *two class periods* of the end of that lab.**

There are two grades for each laboratory:
- your laboratory conduct, execution, and journal grade,

- your formal lab report grade.

Your laboratory journal may be collected and graded at any time. Your lab report will be graded and returned in the next lab session.

Laboratory Class, a Local Scientific Community with Rules for Conduct
- *In all discussions and group work, full respect for all people is required.* All disagreements about work must stand or fall on reasoned arguments about physics principles, data, and/or acceptable procedures, never on the basis of power, loudness, and/or intimidation.

- *It is okay make a reasoned mistake. It is in fact, one of the most efficient ways to learn.* This is an academic laboratory in which to learn, to test your ideas and predictions by collecting data, and to determine which conclusions from the data are acceptable and reasonable to other people and which are not. What is meant by a "reasoned mistake?" It means after careful consideration and a substantial amount of thinking has gone into your ideas, you give your best prediction or explanation as you see it. Of course, there is always the possibility your idea does not accord with the accepted ideas. Then someone says, "No, that is not the way I see it and here is why." Eventually persuasive evidence will be offered for one viewpoint or the other. "Speaking out" your explanations, in writing or vocally is one of the best ways to learn.

- *It is perfectly okay to share information and ideas with colleagues.* Many kinds of help are okay. Since members of this class have highly diverse backgrounds, you are encouraged to help each other and learn from each other. However, **it is never okay to copy the work of others.**

- *Helping others is encouraged because it is one of the best ways for you to learn, but **copying is inappropriate and unacceptable**.* Write out your own calculations and answer questions in your own words. It is okay to make a reasoned mistake; it is wrong to copy. No credit will be given for copied work. It is also subject to school rules about plagiarism and cheating, and may result in dismissal from the course and the school. See the Student-Parent Handbook for further information.

- *Many students use this laboratory each week. Another class probably follows directly after you are done. Respect for the environment and the equipment in the lab is an important part of making this experience a pleasant one.* The lab tables and floors should be clean of any paper or "garbage." Please clean up your area before you leave the lab. The equipment must be either returned to the lab instructor or left neatly at your station, depending on the circumstances.

About Laboratory Equipment

At times equipment in the lab may break or may be found to be broken. If this happens you should inform your instructor. **If equipment appears to be broken in such a way as to cause a danger, DO NOT use the equipment and inform the instructor immediately.**

In summary, the key to making any community work is **RESPECT**.

- *Respect* yourself and your ideas by behaving in a professional manner at all times.

- *Respect* your colleagues (fellow students) and their ideas.

- *Respect* your instructor and his/ her effort to provide you with an environment in which you can learn.

- *Respect* the laboratory equipment so others coming after you in the laboratory will have an appropriate environment in which to learn.

Laboratory Report Writing

Requirements
Each student must submit a laboratory report, which is turned in **two days after the conclusion of each lab activity**, then graded, and returned. The laboratory report must include the following components:

- Title, name, and partners' names,
- Purpose/Problem,
- Hypothesis/Prediction(s),
- Design (if applicable): If the lab has no set procedure, what is to be done? Why is it done this way?
- Discussion/outline of the procedure,
- Data organized in tables, etc…,
- Calculations/Graphs,
- Data analysis,
- Error analysis,
- Conclusion: What was learned and an evaluation of the lab.

Students are required to keep the laboratory reports in an organized notebook. This lab notebook must be kept by students for the entire year and will include the completed lab reports in chronological order. A separate laboratory journal must also be maintained which contains the raw data tables and notes made during the execution of laboratory experiments.

How to Write a Laboratory Report
Many students have a great deal of trouble writing lab reports. They do not know what a lab report is; they don't know how to write one; they don't know what to put in one. This manual seeks to resolve those problems by including examples of an adequate and of an inadequate lab report and examining them in conjunction with this section to aid your understanding.

What Is a Lab Report?
Everyone understands a lab report is a written document about an experiment performed in lab. However, a lab report is **not**:

- a worksheet; you may not simply use the example like a template, substituting what is relevant for your experiment.
- the story of your experiment; although a description of the experimental procedure is necessary and very story-like, this is only one part of the much greater analytical document that is the report.
- rigid; what is appropriate for a report about one experiment may not be appropriate for another.
- a set of independent sections; a lab report should be logically divided, but its structure should be natural, and its prose should flow.

So what is a lab report? A lab report is a document beginning with a question and then proceeds using an experiment to answer that question. It explains not only what was done, but why it was done, and what it means. To try to specify the content in much more detail is too constraining; you must simply do whatever is necessary to accomplish these goals. However, a lab report usually accomplishes them in four phases.

First, it introduces the experiment by placing it in context, usually the motivation for performing it and some question it seeks to answer. Second, it describes the procedures of the experiment. Third, it analyzes the data to yield some scientifically meaningful result. Fourth, it discusses the result, answering the original question and explaining what the result means. There are, of course, other senses of what a lab report is — it is quantitative, it is persuasive, etc...

Now that you have a vague idea of what a lab report contains, consider additional elements.

Making an Argument
A lab report uses an experiment to answer a question, but merely answering it is not enough. The report must convince the reader the answer is correct. This makes a lab report a persuasive document. The persuasive argument is the single most important part of any lab report. You must be able to communicate and demonstrate a clear point. If you can do this well, then your report will be a success; if you cannot, then it will be a failure.

At some point, you have written a traditional, five-paragraph essay. The first paragraph introduces a thesis, the second through fourth defend the thesis, and the fifth paragraph concludes by restating the thesis. This is a little too simple for a lab report, but the basic idea is the same. This structure is typically implemented in science in four basic sections: Introduction, Methodology, Results, and Discussion. This is sometimes called the "IMRD method."

Begin by stating your thesis, along with enough background information to explain it and a brief preview of how you intend to support it in your introduction. Defend your thesis in the methodology and results sections. Restate your thesis, this time with a little more critical evaluation in your discussion. However, keep in mind the IMRD method is a guideline. In this class, we will not have exactly four sections with these titles; we will divide the report more finely. Roughly speaking, "Introduction" will become the Introduction and Prediction sections, "Methodology" and "Results" will become the Procedure, Data, and Analysis sections, and "Discussion" will become the Conclusion section: introduce and state your prediction in the Introduction, Purpose and Prediction sections; test your prediction in the Procedure, Data, and Analysis sections; and restate and critically evaluate both your prediction and your result in your Conclusion section.

Audience
If you are successfully to persuade your audience, you must know something about them. What sorts of things do they know about physics and what sorts of things do they find convincing? For your lab report, they are arbitrary scientifically literate persons. The biggest difference is they do not know what your experiment is, why you are doing it, or what you hope to prove until you tell them. Use physics and mathematics freely in your report, but explain your experiment and analysis in detail.

Technical Style
A lab report is a technical document. This means it is stylistically quite different from other documents you may have written. What characterizes technical writing, at least as far as a lab report is concerned? Here are some of the most prominent features, but for a general idea. Read the adequate sample lab report in this manual.

A lab report does not entertain. When you read the sample reports, you may find them boring; that is okay. The science in your report should be able to stand by itself. If your report needs to be entertaining, then its science is lacking. A lab report is a persuasive document, but it does not express opinions. Your prediction should be expressed as an objective hypothesis, and your experiment and analysis should be a disinterested effort to confirm or deny it. Your result may or may not coincide with your prediction, and your report should support that result objectively.

A lab report is divided into sections. Each section should clearly communicate one aspect of your experiment or analysis. A lab report may use either the active or the passive voice. Use whichever feels natural and accomplishes your intent, but you should be consistent. A lab report presents much of its information with media other than prose. Tables, graphs, diagrams, and equations frequently can communicate far more effectively than words. Integrate them smoothly into your report.

A lab report is usually quantitative. If you do not have numbers to support what you say, you may as well not say it at all. Some of these points are important and sophisticated enough to merit sections of their own.

Nonverbal Media
A picture is worth a thousand words. Take this sentiment to heart when you write your lab report, but do not limit yourself to pictures. Make your point as clearly and tersely as possible; if a graph will do this better than words will, use a graph. When you incorporate these media, you must do so in a way that serves the fundamental purpose of clear communication. Label them "Figure 1" and "Table 2." Give them meaningful captions to inform the reader what information they are presenting. Give them context in the prose of your report. They need to be functional parts of your document's argument, and they need to be well-integrated into the discussion. Students sometimes think they are graded "for the graphs." Avoid these pitfalls by keeping in mind the purpose is communication. If you can make a point more elegantly with these tools, then use them. If you cannot, then stick to tried-and-true prose.

Quantitative Content

A lab report is usually quantitative. Quantitative content is the power of scientific analysis. It is objective. It holds a special power lacking in all other forms of human endeavor: it allows us to know precisely how well we know something. Your report is scientifically valid only insofar as it is quantitative. Give numbers and the numerical errors in those numbers. If you find yourself using words like "big," "small," "close," "similar," etc…, then you are probably not being sufficiently quantitative. Replace vague statements like these with precise, quantitative ones. If there is a single "most important part" to quantitative content, it is error analysis. This lab manual contains a section about error analysis; read it, understand it, and use it.

What Should be My Lab Report?

Introduction

Do three things in your introduction. First, provide enough context so your audience can understand the question your report tries to answer. This typically involves a brief discussion of the hypothetical real-world scenario from the lab manual. Second, clearly state the question. Third, provide a brief statement of how you intend to answer it. It can sometimes help students to think of the introduction as the part justifying your report to your company or funding agency. Leave your reader with an understanding of what your experiment is and why it is important.

Hypothesis/Prediction(s)

Include the same predictions in your report you made prior to the beginning of the experiment. They do not need to be correct. You will do the same amount of work whether they are correct or incorrect, and you will receive far more credit for an incorrect, well refuted prediction than for a correct, poorly supported one. Your prediction will often be an equation or a graph. If so, discuss it in prose.

Procedure

Explain what the actual experimental methodology was in the procedure section. Discuss the apparatus and techniques you used to make your measurements. Exercise a little conservatism and wisdom when deciding what to include in this section. Include all of the information necessary for someone else to repeat the experiment, but only in the important ways. It is important you measured the time for a cart to roll down a ramp through a length of one meter; it is not important who released the cart, how you chose to coordinate the person releasing it with the person timing it, or which one meter of the ramp you used. Omit any obvious steps.

If you performed an experiment using some apparatus, it is obvious you gathered the apparatus at some point. If you measured the current through a circuit, it is obvious you hooked up the wires. One aspect of this which is frequently problematic for students is that a step is not necessarily important or nonobvious just because they find it difficult or time-consuming. Decide what is scientifically important, and then include only that in your report. Students approach this section in more incorrect ways than any other. Do not provide a bulleted list of the equipment. Do not present the procedure as a series of numbered steps. Do not use the second person or the imperative mood. Do not treat this section as though it is more important than the rest of the report. You should rarely make this the longest, most involved section.

Data

This should be the easiest section. Record your empirical measurements here: times, voltages, distance, etc... Do not use this as the report's dumping ground for your raw data. Think about which measurements are important to your experiment and which ones are not. Only include data in processed form. Use tables, graphs, etc... with helpful captions. Do not use long lists of measurements without logical grouping or order. Give the units and uncertainties in all of your measurements. This section is a bit of an exception to the "smoothly integrate figures and tables" rule. Include little to no prose here, most of the discussion belongs in the Analysis section.

Analysis

Do the heavy lifting of your lab report in the Analysis section. Take information from the Data section, scientifically analyze it, and finally answer the question you posed in your Introduction. Do this quantitatively. Your analysis will almost always amount to quantifying the errors in your measurements and in any theoretical calculations you made in the Prediction(s) section. Decide whether the error intervals in your measurements and predictions are compatible. This manual contains a section about error analysis; read it for a description of how to do this. If your prediction turns out to be incorrect, then show that as the first part of your analysis. Propose the correct result and show it is correct in the second part of your analysis. Finally, discuss any shortcomings of your procedure or analysis, such as sources of systematic error for which you did not account, approximations that are not necessarily valid, etc... Decide how badly these shortcomings affect your result. If you cannot confirm your prediction, then estimate which shortcomings are the most important.

Conclusion

Consider your conclusion the wrapping paper and bow of your report. At this point, you should already have said most of the important things, but this is where you collect them in one place. Remind your audience what you did, what your result was, and how it compares to your prediction. Tell them what it means. Leave the reader with a sense of closure. Quote your result from the Analysis section and interpret it in the context of the hypothetical scenario from the introduction. If you determined there were any major shortcomings in your experiment, you may also propose future ways to overcome them.

What Now?

Read the sample reports included in this manual. There are two; one is an example of these instructions implemented well, and the other is an example of these instructions implemented poorly. There is a lot of information here, so using it and actually writing your lab report may seem a little overwhelming. A good technique for getting started is complete your analysis and answer your question before you write your report. At that point, the hard part of the writing should be done: you already know what the question was, what you did to answer it, and what the answer was.

Examples of Adequate and Inadequate Laboratory Reports

Adequate Sample Lab Report

Lab II: Mass and Acceleration of a Falling Ball

Isaac Newton, Lab Partner Galileo Galilei
July 13, 2016

AP Physics 1 Instructor: Albert Einstein

Introduction

The National Park Service is currently designing a spherical canister for dropping payloads of flame-retardant chemicals on forest fires. The canisters are designed to support multiple types of payload, so their masses will vary with the types and quantities of chemicals with which they are loaded. To ensure accurate delivery to the target and desired behavior on impact, the acceleration of the canisters due to gravity must be understood. This experiment therefore seeks to determine the mass dependence of that acceleration. It does so by measuring the accelerations due to gravity of falling balls of several masses.

Prediction

It is predicted that the acceleration of a spherical canister in free fall is mass-independent, as illustrated in Figure 1 on the next page. The acceleration due to gravity of any object near the surface of Earth is assumed to be local g, and there is no reason to expect anything else in these circumstances. Mathematically, $\frac{d\vec{a}}{dm} = \vec{0}$

Procedure

Spherical balls were dropped a height of 1m from rest. Their sizes were approximately the same, and their masses varied from 12.9g to 147.6g. Their free-fall trajectories were recorded with a video camera; Motion Lab analysis software was used to generate (vertical position, time) pairs at each frame in the trajectories and, by linear interpolation, (vertical velocity, time) pairs between each pair of consecutive frames in the trajectories. A known 1-meter length was placed less than 5cm behind the balls' path for calibration of this software. The position and velocity of each ball as functions of time were fit by eye as parabolas and lines, respectively. The acceleration of each was then taken to be the slope of the velocity-versus-time graph, as this was deemed to be more reliably fittable by eye than the quadraticity of the position-versus-time graph.

Data

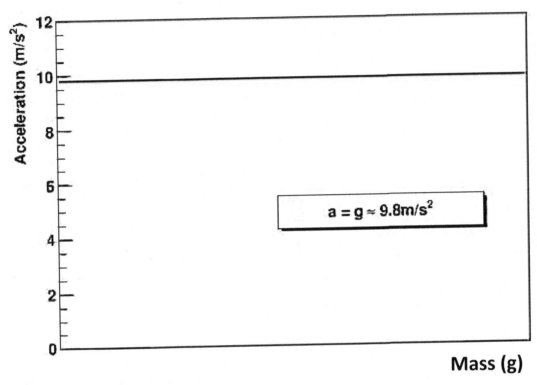

Figure *1*: Magnitude of acceleration due to gravity with respect to mass of a spherical container near Earth's surface; the dependence is predicted to be trivial.

Calculations

M (g)	a (m/s^2)
12.9	9.6
48.8	10.2
55.8	9.8
56.7	9.9
57.7	10.0
143.0	9.7
147.6	9.7

Table *1*: The masses and magnitudes of acceleration of the 7 balls tested in this experiment. The uncertainties in all of the masses are 0.3g. The uncertainties in the accelerations are unknown; see the Analysis section for more information.

Analysis

The accelerations as measured by the velocity fits are given in Table 1 in the Data section. In principle, errors could have been assigned to the fits by finding the maximal and minimal values of the parameters which yield apparently valid fits, but not all groups performed such an analysis, and this group did not have access to the raw data necessary to do so themselves. A method of analysis which does not rely on the errors in the individual accelerations was therefore attempted. In keeping with the hypothesis, the empirical accelerations were treated as independent measurements of local g. A constant was then fit to the data, and the X^2 goodness-of-fit test was used to determine the validity of the hypothesis. The fit is depicted in Figure 2. This yielded a minimal $X^2/\text{NDF} = 0.042$ at $a = (9.84 \pm 0.08)$m/s. The associated p-value is $p = 0.9997$. This suggests the validity of the prediction that the acceleration is mass-independent.

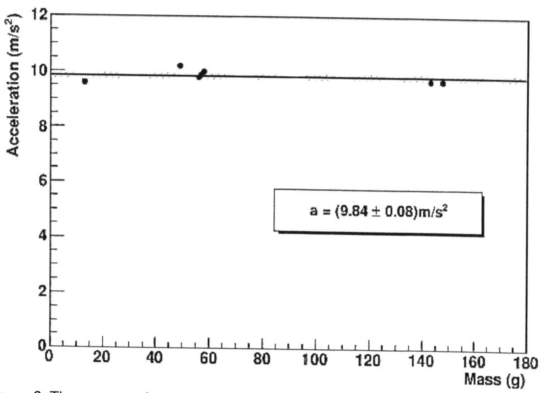

Figure 2: The measured magnitudes of acceleration versus the respective masses, and the constant fit derived therefrom.

Several potentially important sources of error have not yet been addressed. One is the distortion effect of the camera; data was taken only from the center-most portion of the field of view to limit this effect. Another is air resistance; this was assumed to be negligible. Yet another is improper alignment of the calibration object and camera with the balls' trajectories and with one another; this was minimized by the use of a plumb bob. Another is the likely nonzero velocity imparted during release; this was intentionally minimized and then assumed to be negligible. Ultimately, it is not believed that these have significantly affected the result because of the very high p-value of the resulting fit. There is possibly significant systematic error in the mean of the fit acceleration, but the confidence interval is greater than the deviation of this value from the predicted result $(0.08 > |9.81 - 9.84| = 0.03)$, and this does not affect the first derivative, which is constrained to be 0 by the analysis.

Conclusion

Spherical canisters in free-fall were modeled with dropped balls. The mass-independence of the acceleration was confirmed to $p = 0.9997$. This result implies that the National Park Service need not concern themselves with the payload masses of the canisters insofar as gravity is concerned. This result is not to be taken to imply that mass is totally irrelevant, as it may still have significant effects on acceleration due to wind, etc.

Inadequate Sample Lab Report

Lab II
Comte de Rochefort
July 13, 2016

Introduction
We seek to determine how mass affects the acceleration due to gravity of spherical canisters filled with chemicals to fight fires. To do this, we dropped balls from a known height. We used Video Recorder to record videos of them falling, being as careful as possible to simulate the falling canisters accurately and to minimize errors. We analyzed the videos with Motion Lab, taking several data points for each ball.

Prediction

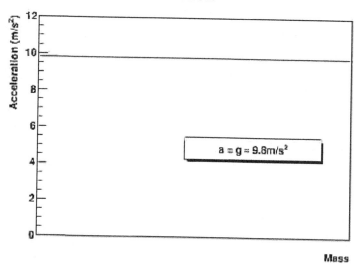

Procedure
We performed this experiment by a scientific procedure. We first made a prediction; then, we performed the experiment; then, we analyzed the data; then, we drew a conclusion. We began by gathering the materials. They included:
- meter stick
- several balls of similar size but different masses
- video camera on tripod
- computer
- tape

We taped the meter stick to the wall for the calibration of Motion Lab. We faced the camera toward the wall. We dropped a street hockey ball with a mass of 57.7g and recorded its video using Video Recorder. We then analyzed its motion using Motion Lab. This began with calibration. We first set time zero at the exact time when we dropped the ball. We then had to calibrate the length. We put the meter stick in the frame of the video, so we used it to do this. We then defined our coordinate system so that the motion of the ball would be straight down.

We then made predictions about the motion. We predicted that the x would not change and that the y would be a parabola opening down with C=-4.9m/s^2. The predicted equations were x(z)=0 and y(z)=-4.9z^2.

We then had to acquire data. We measured the position of the ball at each frame in the video, starting at t=0. We put the red point at the center of the ball each time for consistency. This was important to keep from measuring a length that changed from frame to frame based on where we put the data point on the ball. We also did not use some of the frames at the end of the video, where the ball was at the edge where the camera is susceptible to the fisheye effect and where the ball was not in the frame.

When this was finished, we fit functions to the data points. The functions did not fit the points exactly, but they were acceptably close. We fit x(z)=0 for the x position and y(z)=-5z^2 for the y position. These were close to our predictions.

It then came time to make predictions of the velocity graphs. We predicted that the V_x graph would be a straight line with V_x(z)=0 and that the V_y graph would be a linear line with V_y(z)=-10z.

Next, we fit the functions to the data points for the velocity graphs. We got the predictions exactly right. We then printed our data for the street hockey ball and closed Motion Lab.

We repeated this process for a baseball with a mass of 143.0g. It was mostly the same, with some exceptions. The y(z) fit was y(z)=-4.85z^2 instead of y(z)=-5z^2. The V_y(z) prediction was V_y(z)=-9.7z instead of V_y(z)=-10z. These were also exactly right, so the V_y(z) fit was the same.

At the end of the lab, everybody put their data on the board so we would have enough to do the analysis. We copied it down. Then we were finished, so we started the next experiment.

Data

Ball 1
mass: 12.9+/-0.05g
x prediction: x=0z
x fit: x=0z
y prediction: y=-4.9z^2
y fit: y=-4.8z^2
V_x prediction: V_x=0z
V_x fit: V_x=0z
V_y prediction: V^y=-9.6z

Ball 4
mass: 56.7+/-0.05g
x prediction: x=0z
x fit: x=0z
y prediction: y=-4.9z^2
y fit: y=-4.95z^2
V_x prediction: V_x=0z
V_x fit: V_x=0z
V_y prediction: V_y=-9.9z

V_y fit: V_y=-9.6z

Ball 2
mass: 48.8+/-0.05g
x prediction: x=0z
x fit: x=0z
y prediction: y=-4.9z^2
y fit: y=-5.1z^2
V_x prediction: V_x=0z
V_x fit: V_x=0z
V_y prediction: V_y=-10.2z
V_y fit: V_y=-10.2z

Ball 3
mass: 55.8+/-0.05g
x prediction: x=0z
x fit: x=0z
y prediction: y=-4.9z^2
y fit: y=-4.9z^2
V_x prediction: V_x=0z
V_x fit: V_x=0z
V_y prediction: V_y=-9.8z
V_y fit: V_y=-9.8z

Ball 7
mass: 147.6+/-0.05g
x prediction: x=0z
x fit: x=0z
y prediction: y=-4.9z^2
y fit: y=-4.8z^2
V_x prediction: V_x=0z
V_x fit: V_x=0z
V_y prediction: V_y=-9.6z
V_y fit: V_y=-9.7z

V_y fit: V_y=-9.9z

Ball 5
mass: 57.7+/-0.05g
x prediction: x=0z
x fit: x=0z
y prediction: y=-4.9z^2
y fit: y=-5.0z^2
V_x prediction: V_x=0z
V_x fit: V_x=0z
V_y prediction: V_y=-10.0z
V_y fit: V_y=-10.0z

Ball 6
mass: 143.0+/-0.05g
x prediction: x=0z
x fit: x=0z
y prediction: y=-4.9z^2
y fit: y=-4.85z^2
V_x prediction: V_x=0z
V_x fit: V_x=0z
V_y prediction: V_y=-9.7z
V_y fit: V_y=-9.7z

Analysis
We can calculate the acceleration from the Motion Lab fit functions. To do this, we use the formula $x =x0+v0t+1/2at^2$. Then a is just 2 times the coefficient of z^2 in the position fits. This gives us
Ball 1: a=-9.6
Ball 2: a=-10.2
Ball 3: a=-9.8
Ball 4: a=-9.9
Ball 5: a=-10.0
Ball 6: a=-9.7

Ball 7: a=-9.6

The acceleration can also be calculated using the formula v=v0+at. Then a is just the coefficient of z in the velocity fits. This gives us

Ball 1: a=-9.6
Ball 2: a=-10.2
Ball 3: a=-9.8
Ball 4: a=-9.9
Ball 5: a=-10.0
Ball 6: a=-9.7
Ball 7: a=-9.7

We know the acceleration due to gravity is -9.8m/s^2, so we need to compare the measured values of the acceleration to this number. Looking at the data from the fits, we can see that they are all close to -9.8m/s^2, so the error in this lab must not be significant. Ball 3 actually had 0 error.

We need to analyze the sources of error in the lab to interpret our result. One is human error, which can never be totally eliminated. Another error is the error in Motion Lab. This is obvious because the data points don't lie right on the fit, but are spread out around it. Another error is that the mass balance could only weigh the masses to +/- 0.05g, as shown in the data section. There was error in the fisheye effect of the camera lens. There was air resistance, but we set that to 0, so it is not important.

Conclusion

We predicted that it would be -9.8m/s^2, and we measured seven values of a very close to this. None was off by more than 0.4m/s^2, and one was exactly right. The errors are therefore not significant to our result. We can say that the canisters fall at 9.8m/s^2. This experiment was definitely a success.

Laboratory Grading   Name _____ Period ___

Laboratory Experiment Grading Rubric   Lab Title _____

_____ Has required items, i.e. lab journal, "scientific" calculator, and lab manual (6 points)

_____ Has prediction and method questions completed, if applicable (3 points)

_____ Uses lab equipment appropriately (3 points)

_____ Is focused and on task (3 points)

_____ Stays with lab group (3 points)

_____ Participates with lab group in a helpful manner (3 points)

_____ Treats others and their ideas with respect (3 points)

_____ Maintains appropriate conversational tone and volume (3 points)

_____ No cross-talk with other lab groups (3 points)

_____ **TOTAL** (30 points)

Laboratory Report Grading Rubric

_____ Proper format (per lab manual sample report), punctuation, and spelling (8 points)

_____ Title, name, partners' names, and date (2 points)

_____ Objective/Problem (2 points)

_____ Hypothesis/Prediction (3 points)

_____ Design (if applicable): If no set procedure, what is to be done? Why is it done this way? (10 bonus points)

_____ Discussion or outline of the procedure(s) (5 points)

_____ Data organized in tables, etc... (10 points)

_____ Calculations/Graphs (15 points)

_____ Data analysis (15 points)

_____ Error analysis (5 points)

_____ Conclusion: What was learned and an evaluation of the lab (5 points)

_____ **TOTAL** (70 points)   **GRAND TOTAL** _____ (100 points)

Measurements and Uncertainty

"A measurement result is complete only when accompanied by a quantitative statement of its uncertainty. The uncertainty is required in order to decide if the result is adequate for its intended purpose and to ascertain if it is consistent with other similar results."
National Institute of Standards and Technology

No measuring device can be read to an unlimited number of digits. In addition when we repeat a measurement we often obtain a different value because of changes in conditions we cannot control. We are therefore uncertain of the exact values of measurements. These uncertainties make quantities calculated from such measurements uncertain as well.

Finally, we will be comparing our calculated (experimental) values with an accepted value in order to verify that the physical principles we are studying are correct. Such comparisons come down to the question: Is the difference between your value and the accepted value consistent with the uncertainty in your measurements?

The topic of measurement involves the following concepts:

Sensitivity - The smallest difference that can be read or estimated on a measuring instrument which is generally a fraction of the smallest division appearing on a scale, i.e. about 0.5 mm on a ruler. This results in readings being uncertain by at least this amount.

Variability - Differences in the value of a measured quantity between repeated measurements. It is generally due to uncontrollable changes in conditions such as temperature or initial conditions.

Range - The difference between largest and smallest repeated measurements. Range is a rough measure of variability provided the number of repetitions is large enough. Since range increases with repetitions, we must note the number used.

Uncertainty - How far from the correct value our result may be. Probability theory is needed to make this definition precise, so we use a simplified approach. We take the larger of range and sensitivity as our measure of uncertainty. Example: In measuring the width of a piece of paper torn from a book, we may use a cm ruler with a sensitivity of 0.5 mm (0.05 cm), but find upon 6 repetitions that our measurements range from 15.5 cm to 15.9 cm. Our uncertainty would therefore be 0.4 cm.

Precision - How tightly repeated measurements cluster around their average value. The uncertainty described above is really a measure of precision.

Accuracy - How far the average value may be from the "true" value. A precise value may not be accurate. For example: a stopped clock gives a precise reading, but is rarely accurate. Factors that affect accuracy include how well instruments are calibrated (the correctness of the marked values) and how well the constants in our calculations are known. Accuracy is affected by systematic errors, that is, mistakes that are repeated with each measurement. Example: Measuring from the end of a ruler where the zero position is 1 mm from the end.

Student work should reflect care was taken in all measurement processes. Though the accuracy of results will vary from experiment to experiment, overall error should be relatively small. Percent error is reported as appropriate using the following definition:

% error = (|theoretical value − experimental value| / theoretical value) •100%

Blunders - These are actual mistakes, such as reading an instrument pointer on the wrong scale. They often appear up when measurements are repeated and differences are larger than the known uncertainty. For example: recording an 8 for a 3, or reading the wrong scale on a meter.

Comparison - In order to confirm the physical principles we are learning, we calculate the value of a constant whose value appears in our text. Since our calculated result has an uncertainty, we will also calculate a Uncertainty Ratio (UR) which is defined as:

$$UR = |\text{experimental value} - \text{accepted value}| Uncertainty$$

Uncertainty - A value less than 1 indicates very good agreement, while values greater than 3 indicate disagreement. Intermediate values need more examination. The uncertainty is not a limit, but a measure of when the measured value begins to be less likely. There is always some chance that the many effects that cause the variability will all affect the measurement in the same way. Example: Do the values 900 and 980 agree? If the uncertainty is 100, then UR = 80/100 = 0.8 and they agree, but if the uncertainty is 20 then UR = 80/20 = 4 and they do not agree.

General rule for significant figures
In multiplication and division one needs to count significant figures. These are just the number of digits, starting with the first non-zero digit on the left. For instance: 0.023070 has five significant figures, since one starts with the 2 and count the zero in the middle and at the right. The rule is: Round to the factor or divisor with the fewest significant figures. This can be done either before the multiplication or division, or after.

Example: $7.434 \times 0.26 = 1.93284 = 1.9$ (i.e. 2 significant figures in 0.26).

Reporting uncertainties
There are two methods for reporting a value V, and it's uncertainty U.
- The technical form is "(V ± U) units". Example: A measurement of 7.35 cm with an uncertainty of 0.02 cm would be written as (7.35 ± 0.02) cm. Note the use of parentheses to apply the unit to both parts.
- Commonly, only the significant figures are reported, without an explicit uncertainty. This implies the uncertainty is 1 in the last decimal place.

Example: Reporting a result of 7.35 cm implies ±0.01 cm. Note that writing 7.352786 cm when the uncertainty is really 0.01 cm is wrong.
- A special case arises when there is a situation like 1500±100. Scientific notation allows use of a simplified form, reporting the result as 1.5×10^3. In the case of a much smaller uncertainty, 1500±1, the result is reported as 1.500×10^3, showing the zeros on the right are meaningful.

Additional remarks
- In the technical literature, the uncertainty is also called the error.

- When measured values are in disagreement with standard values, physicists generally look for mistakes (blunders), re-examining their equipment and procedures. Sometimes a single measurement is clearly very different from the others in a set, such as reading the wrong scale on a clock for a single timing. Those values can be ignored, but NOT erased. A note should be written next to any value that is ignored. Given the limited time, it will not always be possible to find a specific cause for disagreement. However, it is useful to calculate at least a preliminary result while still in the laboratory, so you have a chance to find mistakes.

- In adding the absolute values of the fractional uncertainties, the total uncertainty is overestimated since the uncertainties can be either positive or negative. The correct statistical rule is to add the fractional uncertainties in quadrature, i.e.

$$\left(\frac{\Delta y}{y}\right)^2 = \left(\frac{\Delta a}{a}\right)^2 + \left(\frac{\Delta b}{b}\right)^2$$

- The professional method of measuring variation is to use the Standard-Deviation of many repeated measurements. This is the square root of the total squared deviations from the mean, divided by the square root of the number of repetitions. It is also called the Root- Mean-Square error.

- Measurements and the quantities calculated from them usually have units. Where values are tabulated, the units may be written once as part of the label for that column. The units used must appear in order to avoid confusion. There is a big difference between 15 mm, 15 cm, and 15 m.

1. Velocity and Acceleration Lab

Purpose
To study velocity and acceleration by observing a ball's motion on an inclined plane to determine the instantaneous velocity, average velocity, and acceleration of a ball with the inclined plane placed at different angles and distances

Problem
What factors affect the speed and acceleration of a ball? How does changing the angle of an incline influence the velocity and acceleration of a ball rolling down it? What will be the instantaneous velocity, average velocity, and acceleration of a ball rolling down an inclined plane? Will the ball continually accelerate or will it decelerate? Why?

Prediction
If you know the length of an inclined plane and/or runway and the time for a ball going down it, then ... (conclusion)

Equipment
Whiffle golf ball, golf ball, inclined plane, stop watch, meter stick, stack of books, masking tape

Procedure #1
1. Set up an inclined plane at an angle of 10 degrees.

2. Measure the length of the incline plane in meters to the nearest tenth of a centimeter.

3. Place the ball at the top of the incline and allow it to roll down the incline. Record the amount of time required to go from top to bottom.

4. Repeat steps 1, 2, and 3 two(2) more times.

5. Repeat steps one through four for the ball, but change the angle of the incline to 20 degrees.

6. Again, repeat steps one through five for the ball at an angle of 40 degrees.

7. Repeat steps one through seven with a ball of different mass (i.e. whiffle golf ball or golf ball).

Procedure #2
1. Set up a 5m runway with an inclined plane at one end that rises at an angle between 10 and 40 degrees.

2. Place a masking tape mark where the inclined plane touches the floor and label it 0m. Label also 1m, 2m, 3m, 4m, and 5m markers from the bottom of the inclined plane.
3. Take a practice run with the ball. Release it from the top of the inclined plane and begin timing it at the 0m mark.

Data #1

Rolling object – Whiffle golf ball

Incline Angle	Time Trial 1 (s)	Time Trial 2 (s)	Time Trial 3 (s)	Distance of incline plane (m)
10°				
20°				
40°				

Rolling object – Golf ball

Incline Angle	Time Trial 1 (s)	Time Trial2 (s)	Time Trial 3 (s)	Distance of incline plane (m)
10°				
20°				
40°				

Data #2

Pick any angle between 10 and 40 degrees. Take enough runs to get time measurements for the following distances: 0m to 1m, 0m to 2m, 0m to 3m, 0m to 4m, and 0m to 5m. Record the time data.

	Trial 1	Trial 2	Trial 3	Average
0m to 1m	_____	_____	_____	_____
0m to 2m	_____			
0m to 3m	_____	(Student to complete data table)		
0m to 4m	_____			
0m to 5m	_____			

Calculations #1

1. Calculate the average time for all the distances.

2. Calculate the average velocity and acceleration of the ball down the incline by using the formulas $v = d/t$ and $a = V_f - V_o / t$.

Calculations #2

Calculate the instantaneous speed at the following distances:

Average Time (s) (for three trials)	Velocity (m/s) (average of three trials)	Acceleration(m/s²) (average of three trials)

- 1m _____

- 2m _____

- 3m _____

- 4m _____

- 5m _____

Calculate the time between each of the following distances:

- 1m to 2m _____

- 2m to 3m _____

- 3m to 4m _____

- 4m to 5m _____

Average Time (s) (for three trials)	Velocity (m/s) (average of three trials)	Acceleration(m/s²) (average of three trials)

Calculate the acceleration for the following distances:

- 1m to 3m _____

- 2m to 4m _____

- 3m to 5m _____

Graphs #1
1. Make a position versus time graph for your data.

2. Make a velocity versus time graph for your data.

3. Calculate the acceleration of the ball for the data by taking the slope of the velocity versus time graph for these data. Report the acceleration and show your work on your graph. Acceleration is negative for deceleration.

Graphs #2
1. Do a graph of the instantaneous speed and the average speed 0 – 5m. There will be two lines on the graph.

2. Construct a graph for acceleration of the ball at the different angles of the incline. Place the time values for the ball on the x-axis and the velocity values on the y-axis. Remember all blocks (squares) on each axis must have an equal value when plotting the data.

Conclusion
Questions to consider in the discussion of results:

1. What are the shapes of the lines?

2. Which line shows greater acceleration?

3. Did your ball travel at a constant speed? How do you know?

4. How could you change the experiment to make the ball decelerate faster?

5. How could you change the experiment to make the ball accelerate faster?

6. How can you change the experiment to make the ball not accelerate or decelerate for an entire 5m?

7. What is happening to the speed of the ball as it continues down the inclined plane? Use your data to support your answer.

8. What are some factors that may introduce error into this experiment?

9. What would a distance versus time graph look like if you started taking data at the top of the inclined plane and continued after it reached the bottom?

10. How did changing the angle of the inclined plane change the velocity and acceleration of the ball?

11. Was the graph you constructed for acceleration linear in nature?   Why or why did not it take this shape?

12. What force(s) caused the ball to roll down the incline?

13. Why was it important to do multiple trials?

2.a Visual Reaction Time Lab

Purpose
To understand the role reaction time plays in the gathering of data in experimental procedures

Prediction
Have you ever wondered what your visual reaction time is? What do you think your visual reaction time is? Make a prediction.

Procedure
Can your group find a method for measuring a person's visual reaction time? Write the method as a list of procedures numbered 1 to whatever. Put these procedures in a lab report in a section titled "Procedures for Determining Reaction Time."

Equipment and Measurements
Once you have a set of procedures, you should know what quantities you need to measure and what tools you need to measure it, and how these tools will be used. Put this information in a section titled "Quantities to be Measured and Tool(s) Required."

Data
If your procedures are correctly written, you should know what data to collect and how to collect the data. Make a data table. Title this section "Data Tables."

Calculations
Using the data collected, calculate reaction time. Write these calculations in a section titled "Calculations of Reaction Time."   The student whose data are being used must be indicated.

1. If traveling at 88km/hr how far do you go in a time equal to the lab reaction time?

2. Convert the distance to miles, and then convert to feet.

3. If driving at 32m/s how far would you go during the reaction time?

4. Convert the distance to miles, then convert to feet.

5. Find Internet information on how alcohol affects a person's reaction time. Using this information, recalculate #s 1-4. Provide the URL so I may confirm your sources!

6. Find Internet information on how texting while driving affects a person's reaction time. Using this information, recalculate #s 1-4. Provide the URL so I may confirm your sources!

Conclusion

Provide a summary, which includes the prediction of each student. The summary should start with a brief description of what your group did and how you did it. Then, each summary must have a comment comparing the predicted reaction time and the actual calculated reaction time. How could you find a person's auditory reaction time? Finally, write a paragraph telling what you learned from this lab.

2.b Visual Reaction Time Home Lab

Purpose
To understand the role reaction time plays in the gathering of data in experimental procedures

Prediction
Have you ever wondered what your visual reaction time is? The time it takes you to see, process the movement, and react. What do you think your visual reaction time is? Make a prediction.

Procedure
Can you find a method for measuring your reaction time? Write your method as a list of procedures numbered 1 to whatever. Put these procedures in a report in a section titled "Procedures for Determining Visual Reaction Time."

Equipment and Measurements
Once you have a set of procedures, determine what quantities you need to measure, what tool(s) you need to measure it, and how the tool(s) will be used. Put this information in a section titled "Quantities to be Measured and Tool(s) Required."

Data
If your procedures are properly written, you know what data to collect and how to collect it. Make a data table. Title this section "Data Table."

Calculations
Using the data collected, calculate your reaction time. Write these calculations in a section titled "Calculation of Reaction Time."

1. If traveling at 70 mi/hr how far do you go in a time equal to your reaction time? Convert it to feet (ft).

2. Find Internet information on how alcohol affects a person's reaction time. Using this information, then recalculate # 1. Provide the URL so I can confirm your sources.

3. Find Internet information on how texting while driving affects a person's reaction time. Using this information, then recalculate # 1. Provide the URL so I can confirm your sources.

Conclusion
Provide a summary, which starts with a brief description of what you did and how you did it. Compare the predicted reaction time and the actual calculated reaction time. How could you find a person's auditory reaction time? Write a paragraph telling what you learned from this lab.

3. Projectile Motion Lab with a Horizontal Launch

Purpose
To determine if Newtonian kinematics predicts the motion of a horizontally launched projectile

Discussion
In this lab, you will check if the kinematic concepts and equations discussed in class predict the motion of a projectile.

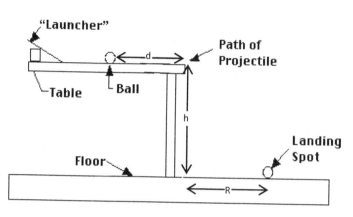

You will measure the starting velocity of a projectile and the distance from the table (range, R) that the projectile lands. From this, you will calculate the vertical distance it fell and compare your calculated value to the measured value.

Record the location where the projectile lands on the floor by placing a piece of carbon paper carbon side down over a piece of paper taped to the floor. The projectile will leave a mark on the paper where it lands. You can measure the horizontal range of the projectile and compare this to the calculated distance.

Equipment
Ball, meter stick, stopwatch, string, carbon paper, incline track, tape, small weight, white 8.5 x 11 inch paper

Procedure
1. Devise a "launcher" for the ball from an inclined track. Find a starting point on the launcher that gives the ball a reasonable velocity.

2. Place two (2) pieces of tape 30 to 50 cm apart on the table in the path of the ball. This is distance "d" in the diagram above. Record this distance. The distance you use needs to be a compromise:
   o If the distance is too short, you will not be able to get an accurate time for the ball to cover the distance and the velocity will not be accurate.
   o If the distance is too long, friction will slow the ball appreciably by the time it reaches the edge of the table and the calculated speed will not be the actual speed of the ball when it leaves the table.

3. Carefully measure the vertical distance from the top of the table to the floor. This is "h" in the diagram above. Record the distance.

4. From a trial run, find the approximate position where the projectile hits the floor. Tape a piece of white paper at this location and put the carbon paper face down over it to identify the landing spot of the projectile.

5. Launch the projectile at least three (3) trials.
   - For each launch, measure the time it takes the ball to roll the measured horizontal distance on the table and record the rolling times (t_{roll}) in a data table.
   - You should get a group of reasonably close together spots on the "target" paper. If the spots are wildly far apart, you need to adjust your launcher or launching technique to get more consistency.

Locate the point on the floor directly below the edge of the table top where the ball leaves the table. You can do this accurately by making a "plumb line" from a small weight and a string. Measure the distance from this point to the center (average) of your landing positions. This is the range of the projectile ("R" in the diagram).

Change your launcher's angle to the tabletop so the projectile is launched at a different angle and repeat steps 4 to 6. Do at least three (3) different angle trials.

Calculations

For the data collected at each angle:

1. Calculate the average rolling time (t_{roll}) for the projectile to travel the measured horizontal distance (d).

2. Calculate the speed, v_x, of the projectile as it rolls across the table ($v_x = d/t_{roll}$). This should be approximately the speed the projectile has when it leaves the table.

3. Calculate the time (falling time, t_{fall}) it took the projectile to move the horizontal distance (range) R. (Since $R = v_x t_{fall}$, $t_{fall} = R/v_x$)

4. Calculate the distance the ball will fall vertically from the table top to the target.
$$h = \frac{1}{2}gt_{fall}^2$$

5. A good measure of comparison (between the measured and calculated heights) is the percent of difference:
$$\% \text{ difference} = \frac{\text{difference in values}}{\text{accepted values}} \times 100\%$$

Conclusions

1. How do the measured height and calculated height compare?

2. Do the kinematics equations seem to work in practice? If not, why not?

3. Do you think it is the fault of the kinematics or due to a problem with experimental conditions and/or procedure? Be specific.

4. Projectile Motion Lab with Launch at an Angle

Purpose
To examine the factors effecting trajectories at an angle

Equipment
Launcher, plastic ball, meter stick, stopwatch

Procedure
1. Set the launcher at 60°, 45° and 30° above the horizontal.

2. Carefully plunge the plastic ball into the barrel to the full setting.

3. Launch the ball, allowing it to bounce several times while observing, noting the path.

4. Launch the ball and record the total time of flight for three (3) flights and the horizontal distance d_x the ball travels at the first bounce. *Take the average time and average distance.*

Data
Construct and record data in a table.

Calculations

1. Using the motion equations for v_y, v_x, d_y, d_x determine the:
   - vertical (v_y) velocity of the ball,
   - horizontal (v_x) velocity of the ball,
   - maximum height (d_y) of the ball.

2. Establish a scale of 1cm = 0.5 m/s, then do a graphical vector analysis to determine the initial velocity (v_o). Verify it using the Pythagorean Theorem.

3. Determine the following for your launch:
   - g at the maximum height (d_y) of the ball and 0.25 meters from the horizontal surface,
   - velocity of the ball upon impact with the floor,
   - vertical and horizontal velocities at the top (apex) of the flight.

4. On one graph, sketch the v_y and v_x of the ball for the total time of flight. What is the significance of the slopes? What is the significance of the area under the v_x velocity graph?

5. Atwood Machine Lab

Purpose
To determine the relationship between the factors that influence the acceleration of an Atwood's machine demonstrating the consistency of Newton's Laws of Motion and to measure the acceleration due to gravity

Theory

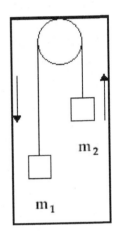

A simple Atwood's machine is shown. An Atwood's machine consists of two masses connected by a light (taken to be massless) string running over a light (massless) pulley suspended above a floor or table. If one of the masses is greater than the other (e.g., $m_1 > m_2$), the system moves as shown. If one assumes a light pulley and light string which does not stretch, this system lends itself to a relatively simple analysis using Newton's Laws. You will analyze data from an Atwood machine and compare your predicted values with those derived from observing the physical behavior of an actual Atwood's machine.

Description
When released the heavier mass accelerates downward while the lighter one accelerates upward at the same rate. The acceleration depends on the difference between the two masses, as well as the total mass. For $m_2 > m_1$, the acceleration is given by $a = g (m_2 - m_1)/(m_2 + m_1)$. By picking the value of the masses, the acceleration can be tuned to a manageable value so the elapsed times may be measured accurately by hand. Using the elapsed time, then you can calculate "a" and in turn "g."

Analysis
Begin by constructing free body diagrams for each of the masses in the figure above. By assuming the pulley is massless, you are not concerned about its effect in free body diagrams other than redirecting a force. The only thing the single massless pulley does is couple the two masses with a cord that has the same tension (T) throughout it.

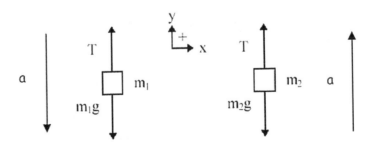

Assuming that up and right are positive directions and using Newton's second law in the y direction (n. b. there are no forces in x):

$$\sum F_y = ma \qquad (1)$$

$$T - m_1 g = -m_1 a \qquad (2)$$

$$T - m_2 g = m_2 a \qquad (3)$$

Notice the only effect the pulley has on this system is to redirect "T" from one mass to the next. Since T is the same everywhere in this system, (2) and (3) are set equal to each other:

$$m_1 g - m_1 a = m_2 g + m_2 a$$

then, rearranging terms:

$$m_1 g - m_2 g = m_1 a + m_2 a$$

$$(m_1 - m_2)g = (m_1 + m_2)a$$

$$a = \frac{m_1 - m_2}{m_1 + m_2} g \qquad (4)$$

If one is able to measure the distance either mass moves (the same distance, in one case up, in the other case, down) starting from rest ($v_0 = 0$) to the point just before m_1 hits the ground, the final speed of the system (the common speed of both masses just before m_1 hits the ground) may be calculated as a function of acceleration and distance "h":

$$v^2 = v_o^2 + 2a(y - y_o) \rightarrow v^2 = v_o^2 + 2ah \rightarrow v = \sqrt{2ah}$$

Finally, substituting the value for the acceleration "a", that was obtained from equation (4), the final speed of the system is obtained:

$$v = \sqrt{2 \frac{m_1 - m_2}{m_1 + m_2} gh}$$

Equipment
Meter stick, stopwatch, masses, Atwood's machine pulley, string

Procedure
1. Create a data table to record your data and subsequent calculations. Use masses such that the drop time is large enough that your reaction time is minimized.

Trial	Mass 1 (g)	Mass 2 (g)	Time (s)	Acceleration (m/s²) (experimental)	$\frac{M1 - M2}{M1 + M2}$	Acceleration (m/s²) (theoretical)	Gravitational Constant (m/s²) (experimental)

2. Set up your Atwood's machine. Mount the pulley as high as possible. Tie one end of the string to mass 1 and tie the other end to mass 2. Hang the masses over the pulley, then adjust the pulley height so the bottom of the upper mass is 55-65 cm above the floor when the lower mass is on the floor.

3. Pull the lighter side to the floor. Measure the time required for the heavier side to hit the floor. Drop the mass **three times**, and then **repeat for five mass pairs**. Each drop should have its own row in the data table. The displacement "d" and average elapsed time "t" are related by the following equation $d = \frac{1}{2} a t^2$. Thus, you can find acceleration (experimental) from this equation. It is good experimental procedure to check the quality of your data as you go. Calculate a value for "g" for each drop. If there are large deviations from the expected value of "g," discuss your experimental procedures with your lab partners.

Data Analysis

Make a graph of your data with acceleration on the y-axis and $(m_2 - m_1)/(m_1 + m_2)$ on the x-axis. Fit a line to these data. The slope should be the acceleration due to gravity, and the intercept should be close to zero. You can also choose to set it to zero when you do the line of fit. Include the graph in your report. The slope of the line is one way to determine a value for "g" and it is the preferred method. Another way of determining "g" is to take the average of the experimental values of the gravitational constant. Does the difference between the two values fall within the expected error? If the value from the slope is much different, then the "point values," this usually indicates a source of error. Discussion of this should be found in your conclusion.

Conclusion

1. Compare your experimental value of "g" to the expected value. A good measure of comparison (between the calculated and accepted gravitational constant) is the percent of difference:

$$\% \text{ difference} = \frac{\text{difference in values}}{\text{accepted values}} \text{ x } 100\%$$

2. What sources of errors may have affected your result? Discuss the accuracy of your results and comment on any deviation from the expected value.

3. Is a simple Atwood's machine a mechanically conservative or a non-conservative system?

4. How would you redesign the experiment to provide greater accuracy? Not lighter pulleys or strings, or better measuring devices, but something in the basic adjustable parameters of length, mass, time, etc...

5. Explain the shape of the $(m_2 - m_1)/(m_1 + m_2)$ vs. acceleration graph. What does it indicate about acceleration?

Alternative Lab

Atwood Machine (simulation)

Read the Purpose, Theory, Description, and Analysis of the prior Atwood Machine Lab.

Procedure
Go to: http://www.wiley.com/college/halliday/0470469080/simulations/sim20/sim20.html

Data

Trial	Mass 1 (kg)	Mass 2 (kg)	Time (s)	M_2-M_1 (kg)	M_1+M_2 (kg)
1					
to					
11					

Calculations
For each trial, calculate the:
- Distance each mass fell, and
- Experimental velocity.

Graph
Make a graph of your data with acceleration on the y-axis and $(m_2- m_1)/(m_1 + m_2)$ on the x-axis. Fit a line to these data. The slope should be the acceleration due to gravity, and the intercept should be close to zero. You can also choose to set it to zero when you do the line of fit. Include the graph in your report. The slope of the line is one way to determine a value for "g" and it is the preferred method.

Conclusion
1. What sources of errors may have affected your result? Discuss the accuracy of your results and comment on any deviation from the expected value. The percent error you arrived at should be very small.

2. Explain the shape of the $(m_2- m_1)/(m_1 + m_2)$ vs. acceleration graph. What does it indicate about acceleration?

6. Incline Plane Forces and Friction Lab

Purpose
To investigate incline plane forces and friction and measure the static and kinetic coefficients of friction

Equipment
Spring scale (in newtons), block of wood with two different surfaces, wooden track, string, protractor

Procedure
1. Tie the string to the end of the object. Hang the object by the string from the spring scale. Measure the weight of the object in newtons.

2. Place the wooden track on a horizontal surface. Hold the spring scale, and with the string held parallel to the level track, pull the object along the track at a constant speed. With the spring scale, measure the amount of force required to keep the object moving at a uniform rate. Repeat this procedure several times (minimum of 3 times), average your results – this is the value for the force of kinetic friction between the surface of the track and the surface of the object.

3. Detach the spring scale from the object and place the object on the wooden track. Slowly lift one end of the track. Continue increasing the angle of the track with the horizontal until the object starts to slide. Use a protractor to measure this angle. Record the value of this angle as the angle of static friction. The tangent of this angle is the coefficient of static friction.

4. Move the object to one end of the track. Again, slowly lift this end of the track while your lab partner lightly taps the object. Adjust the angle of the track until the object slides at a constant speed after it has received an initial light tap. Use the protractor to measure this angle and record it as the angle for kinetic (sliding) friction. The tangent of this angle is the coefficient of kinetic (sliding) friction.

5. Repeat the experiment with the other surface of the block facing downward.

Data
Construct and record data in a table.

Calculations
Calculate the coefficients of static and kinetic friction for each surface used. Show your work and compare/contrast the two coefficients. Explain the difference.

$$\mu = \frac{F_f}{F_\perp} = \frac{F_\parallel}{F_\perp} = \frac{F_w \sin\theta}{F_w \cos\theta} = \tan\theta.$$

Questions

1. Throughout this experiment, the string should be parallel to the surface of the plane. Why is this important?

2. Explain any differences between the values for the coefficients of static and kinetic friction.

3. Using the average force of kinetic friction from the data, calculate the coefficient of kinetic friction for each surface. Show your work and compare it to the coefficient that was calculated using the angle – finding a percent difference. Explain the difference.

4. A brick is positioned first with its largest surface in contact with an inclined plane. The plane is tilted at an angle to the horizontal until the brick just begins to slide. The angle, θ, of the plane with the horizontal is measured. Then the brick is turned on one of its narrow sides, the plane is tilted, and θ is again measured. Predict whether there will be a difference in these measured angles. Explain your answer in terms of the equation for the force of friction. Is the coefficient of static friction affected by the area of contact between the surfaces?

5. A brick is placed on an inclined plane tilted at an angle to the horizontal until the brick just begins to slide. The angle, θ, of the plane with the horizontal is measured. Then the brick is wrapped in waxed paper and placed on a plane. The plane is tilted, and θ is again measured. Predict whether there will be a difference in these measured angles. Explain.

Physics Lab Manual

6. From the previous two answers, determine the factors that influence the force of friction.

7. While looking for new tires for your car, you find an advertisement offering two brands of tires, brand X and brand Y, at the same price. Brand X has a coefficient of friction on dry pavement of 0.90 and on wet pavement of 0.15. Brand Y has a coefficient of friction on dry pavement of 0.88 and on wet pavement of 0.45. If you live in an area with high levels of precipitation, which tire gives you better traction? Explain.

7. Galileo Ramps Lab

Introductory Video
http://www.pbslearningmedia.org/resource/phy03.sci.phys.mfw.galileoplane/galileos-inclined-plane/

Discussion
Galileo Galilei was a physicist, astronomer, mathematician, and creative mastermind who lived in the 16th and 17th centuries in Italy. He was the inventor of the telescope, and one of the first people to suggest the Earth traveled around the Sun and not the other way around. He was very interested in physics and how things worked on Earth, and he conducted a lot of experiments to observe gravity and natural phenomena, before they were mathematically described by Sir Isaac Newton.

Galileo began experimenting with falling objects and testing the idea that even though objects have different masses, they fall towards the Earth at the same velocity. Because of timing and other factors like wind resistance are an issue at great heights (like dropping a ball from the height of a building), Galileo used inclined planes, i.e. ramps to conduct his experiments. Galileo stated objects in a vacuum fall to the Earth with a constant acceleration.

Purpose
To determine the acceleration due to gravity on Earth

Equipment
Inclined plane, books to stack, meter stick, protractor, golf ball, stopwatch

Procedure
1. Stack some books and set one side of the incline plane on the books to create a ramp.

2. Use the protractor to measure the angle between the ramp and the floor. Adjust the stack of books until you can get the ramp as close to 30° as possible. Record the final angle.

3. Use the meter stick to mark 10 cm intervals along the ramp, starting at the floor and going upward.

4. Set the golf ball at a measured distance along the ramp. Time how long it takes for the golf ball to hit the floor after you release the ball. Record both the distance at which you let the ball go and the time it takes the ball to travel the length of the ramp.

5. Repeat step for at different lengths along the ramp.

6. Graph your results. Put time on the x-axis, and distance traveled on the y-axis. Do you notice any patterns?

7. Calculate the acceleration for the points you tested using the equation $a = 2d / t^2$

8. Try the experiment again with at least two (2) different ramp angles.

Results
The acceleration at each point should be almost the same. Differences can be connected to imperfections in timing and friction on the ramp.

Conclusion
The graph you create will show the longer the ball is on the ramp, the faster it will move. With constant acceleration, the velocity of an object will get increasingly faster. The constant acceleration in the experiment is due to gravity. Acceleration due to gravity is measured as 9.81 m/s². You will not measure this acceleration because of the inclined plane, but if you were to conduct an experiment by dropping balls from different heights, this is what you would expect. If you change the angle of the ramp to be steeper, will the acceleration you record will be closer to that of gravity?

8. Kepler's Laws Lab

Purpose

To become more familiar with Kepler's Laws of Planetary Motion (This activity was modified from the Genesis Mission *Search for Origins* education series.)

http://astro.unl.edu/naap/pos/animations/   or

http://phet.colorado.edu/sims/my-solar-system/my-solar-system_en.html

Equipment

Cardboard, two (2) push pins, string, pencil, calculator, one of the animation links above, computer with Internet, white 8.5" X 11" paper, ruler

Procedure

Part 1: Drawing an Ellipse and Calculating Eccentricity-Kepler's First Law of Planetary Motion

1. Tie your piece of string in a loop.

2. Place your paper on the cardboard and put your pushpins in the middle of the page lengthwise. The pushpins should be about 10 centimeters apart.

3. Changing this distance will change the shape of your ellipse.

4. Put your loop of string over the ends of the pushpins. Draw the loop tight with the tip of your pencil and form a triangle with your string. Keep the loop tight and draw an ellipse.

5. Remove the string and push pins from your paper.

6. Label each hole made by the push pins "focus 1" and "focus 2."

7. Choose one of these foci and label it "Sun."

8. Choose a place on the outline of your ellipse and place a dot there. Label the dot with a planet name of your choosing. (e.g. Planet Hepnery)

9. Find the point on the outline of the ellipse closest to the dot you made the Sun. Label this point "Perihelion."

10. Find the point on the outline of the ellipse farthest from the dot you made the Sun. Label this point "Aphelion."

11. Put an "X" directly in the center of your ellipse exactly half way between the two foci.

12. Draw a line from the "X" to the dot you denoted as the Sun. Label this line as "c."

13. Draw another line from the "X" through the focus that does not denote the Sun and all the way to the point you denoted "Aphelion." Label this line as "a." This line is the "semi-major axis." It is similar to the radius of a circle.

14. Eccentricity is the measurement of how stretched out an ellipse is. It ranges from zero to one. Zero is the eccentricity of a circle and one is the eccentricity of a straight line. Calculate the value of the eccentricity for the ellipse you drew by measuring the length of line "c" and measuring the length of line "a." Calculate the eccentricity of the ellipse by taking "c" and dividing it by "a." Put the data in your data table.

Length of Line "c" (cm)	Length of Line "a" (cm)	Eccentricity of the Ellipse (c/a)

After doing this activity, what does Kepler's First Law of Planetary Motion say?

Staple the ellipse drawing you made to your lab report when you submit it.

Part 2: Calculating the Eccentricity of Planet Orbits
1. Calculate the eccentricity of each planet by using the formula $e = c/a$. State your answer in the proper number of significant figures.

Planet	Distance from Center of Ellipse to Focus in Astronomical Units (c)	Semi-Major Axis in Astronomical Units (a)	Eccentricity (e)
Mercury	0.080	0.387	
Venus	0.005	0.723	
Earth	0.017	1.000	
Mars	0.142	1.524	
Jupiter	0.250	5.203	
Saturn	0.534	9.540	
Uranus	0.901	19.180	
Neptune	0.271	30.060	
Pluto	9.821	39.440	

2. Which of the planet's orbits is the most eccentric? Assume Pluto is still a planet for this question.

3. Which of the planet's orbits is the least eccentric (closest to a circle's eccentricity of zero)? Assume Pluto is still a planet for this question.

4. Which two planets have the most similar eccentricity?

5. Which planet has an eccentricity most similar to Earth's eccentricity?

6. The average eccentricity of the Moon's orbit around the Earth is 0.054900489. Would you say the eccentricity of the Moon's orbit is low, medium, or high with respect to most of the planets' orbits around the Sun?

7. How could the eccentricity of a planet's orbit affect the amount of solar radiation it receives from the Sun?

Part 3: Kepler's Second Law of Planetary Motion

1. Go to one of the animation links on page 49. Set up an orbit of a planet around the Sun that is fairly elliptical by adjusting the mass and/or velocity of the orbit. Run the animation.

2. How does the speed of a planet's orbit at perihelion compare to the speed of a planet's orbit at aphelion? Why is there a difference in speed?

3. Look at the diagram below. **Count** the approximate number of squares in sector 1 and sector 2.

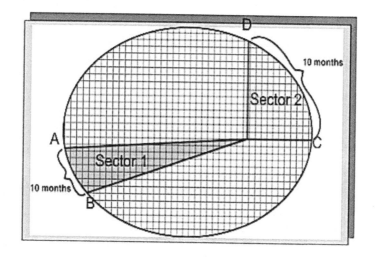

Squares in Sector 1: _____

Squares in Sector 2: _____

4. What can you say about the number of squares in Sector 1 compared to the number of squares in Sector 2? What does the number of squares imply about each sector's area?

5. If it takes the same amount of time for a planet to move from point A to point B as it does for a planet to move from point C to point D, then what must a planet do in terms of its speed in each sector? Note that the distance between A and B is shorter than the distance between C and D.

Speed from A to B (Faster or Slower?)	Speed from C to D (Faster or Slower?)

6. Based on what you have seen, Kepler's Second Law says planets sweep out equal _____ in equal _____. To do this, planets _____ when closer to the Sun and they _____ when farther from the Sun.

7. Earth's perihelion is in January and its aphelion is in July. Why is this not the reason for the seasons on Earth? If it was, the Northern Hemisphere on Earth would be hotter in January and colder in July.

Part 4: Kepler's Third Law of Planetary Motion
Use the following chart to answer the questions that follow.

Planet	Mean Orbital Velocity and Mean Planet Distance to the Sun								
	Mercury	Venus	Earth	Mars	Jupiter	Saturn	Uranus	Neptune	Pluto
Mean Orbit Velocity in (km/s)	47.87	35.02	29.79	24.13	13.07	9.67	6.84	5.48	4.75
Mean Distance to the Sun in Astronomical Units (AU)	0.39	0.72	1.00	1.52	5.20	9.54	19.19	30.07	39.48

1. How does a planet's distance from the Sun affect the planet's orbital velocity? In other words, do planets that are farther from the Sun travel faster or do they travel slower?

2. Based on your response to number 1, what does Kepler's Third Law of Planetary Motion say?

Conclusion
How might Kepler's Laws be used by the Jet Propulsion Laboratory (JPL) to plan missions to other planets in terms of timing the mission launches? If you have time, go back to the computer animation link and play with it to see how crazy planetary orbits can get!

9. Hooke's Law Lab

Purpose
To find the spring constant for three different springs, use the constants to find the mass of objects of unknown mass, and to explore how the equivalent spring constant changes when springs are added in series and in parallel

Prediction
From the description of the way a spring stretches when a mass hangs from it, predict what the graph would look like of force (mg) as a function of stretch (Δx). Draw it in your lab journal.

Equipment
Ruler and pole arrangement; three different springs; masses from a few grams to several hundred grams

Procedure

Part 1: Find the Spring Constant (k) of Three Different Springs
1. First, hang a spring on the pole, and mark the height of the bottom of the spring without anything hanging from it.

2. Next, hang a small mass from the bottom of the spring, and note the Δ X, i.e., the amount the spring stretched.

3. Add mass in constant increments. For a stiffer spring, go in larger increments. Record the amount of stretch relative to the initial position. Record your data in a table.

Mass (kg)	Weight (n)	X (m)	$\Delta x = x - x_1$ (m)

4. Graph force (mg) as a function of stretch (Δx), i.e. on the y-axis you have force (mg) and on the x-axis you have stretch (Δx).

5. Repeat for two more springs. Hint: Partner with another group, and each of you take one spring from the mixed spring box. Be sure you each have springs with different "stretchiness." Each group gets the data table and graph for one spring, and then trade. In this way, you will have three different graphs, but you save time.

You now have three graphs for three different springs. Calculate the spring constant for each spring from each graph.

$k_1 =$ _____ $k_2 =$ _____ $k_3 =$ _____

Part 2: Find the Mass of Three Unknown Objects using Your Graph

1. Select three objects the weight of which you do not know, such as someone's keys, a shoe, a book, etc… Hang each object, one at a time, from one of your springs and record the amount of stretch.

2. Plot this stretch on the graph for that spring on the x-axis, and using the graph, find the weight of the object by reading up from the x-axis to the line $mg = kx$, and then reading across to the weight.

3. Weigh your object and compare this measurement with the weight you derived from your graph. How close were you?

4. You should have all three graphs in your lab journal, as well as the three data tables, and the masses of your unknowns from your graphs and the comparison with the weight from the triple beam balance.

Part 3: Find the Equivalent "k" for Springs in Series and in Parallel

Prediction: Do you think it takes more force to stretch three springs hooked together in series (end-to-end) or three springs hooked together in parallel (next to each other)? Write your prediction, and explain your reasoning.

After you write a prediction, work with another group and three identical springs. Try putting them next to each other and end-to-end and describe what you see.

1. Hang three springs that seem to be identical, side-by-side from the horizontal portion of the pole. Put a light stick across their bottoms, such as a lightweight plastic pen. Neglect the masses of the springs and the pen. Note the position of the bottom of this parallel arrangement of springs on the pole.

2. Hang three(3) to five(5) different masses from this parallel arrangement, and record the displacement of the system for each weight in a data table, just as in Part 1.

3. Graph these data, as in Part 1, with the force on the vertical axis and the displacement (Δx) on the x-axis. Calculate "k-equivalent" for this system from the graph.

4. Is k-equivalent greater, the same as, or less than, k for an individual spring?

5. Now hook the three springs end-to-end, i.e. series. What do you notice?

6. Note the position of the bottom of each spring, with no extra mass on the bottom, and record these in a data table ($x_{1\text{-initial}}$, $x_{2\text{-initial}}$, and $x_{3\text{-initial}}$) and the initial position of the bottom of the whole thing (which is also $x_{3\text{-initial}}$).

7. Now hang a mass from the bottom, and note the new positions of the bottom of each spring (Δx_1, Δx_2 and Δx_3) and the total displacement, Δx_{total}). Repeat this for three(3) to five(5) masses, record in a data table, and graph force (mg) as a function of displacement of the system (Δx_{total}). Find k-equivalent for this system from the graph.

8. Is the k-equivalent for three springs in series the same as, greater than, or less than, the k for any of these springs alone?

9. When you add springs in parallel, is the equivalent spring constant of the system greater than, same as, or less than the spring constant of one spring alone?

10. When you add springs in series, is the equivalent spring constant of the system greater than, less than, or the same as the spring constant of one spring alone?

11. Derive the expression for finding k-equivalent for springs in parallel and in series. You may find yourself having to build something with springs one day and for another, when you study electrical circuits, you can use springs as the mechanical analog of certain parts in the circuit. This can be convenient for modeling the behavior of systems!

10. Pendulum Properties Lab

Purpose
To determine the factors affecting the swing of a pendulum, such as mass, height to which it is raised, and length of the string

Equipment
Ring stand, support arm, string, masses, stopwatch, protractor

Procedure
1. Attach a string to the support arm of the ring stand. At the end of the string, attach a mass. Then release the string and mass, letting it swing at an angle of 15°. Time how long it takes for the mass to return to the starting point. After recording the data, replace the original mass with a different mass. Repeat two more times.

2. Test the time a constant mass attached to a string takes to swing and return to its original position at 5°. Then raise the height of release to 10° and then to 15°.

3. Test the time a constant mass attached to a string takes to swing from a point and back to its original position, for three different string lengths.

Data
Make a data table.

Conclusion
After completing the three experiments above, analyze your data and answer the following:

1. What factors affect the time of a pendulum's swing?

2. How does the amount of mass affect the time?

3. How does the height of release affect the time?

4. How does the length of the string affect the time?

Justify your conclusions using your data.

11. Momentum and Collisions Lab

Purpose
- To draw "before-and-after" pictures of collisions
- To apply law of conservation of momentum to solve problems of collisions
- To explain why energy is not conserved and varies in some collisions
- To determine the change in mechanical energy in collisions of varying "elasticity"
- To determine what "elasticity" means

Equipment
Go to: http://phet.colorado.edu/en/simulation/collision-lab

Procedure
Record necessary information and data in your lab journal and answer the following questions in your lab report:

1. In the green box on the right side of the screen, select the following settings: one dimension, velocity vectors ON, momentum vectors ON, reflecting borders ON, momenta diagram ON, elasticity 0%. Look at the red and green balls on the screen and the vectors that represent their motion.
   - Which ball has the greater velocity?
   - Which has the greater momentum?

2. Explain why the green ball has more momentum but less velocity than the red ball (HINT; What is the definition of momentum?).

3. Push "play" and let the balls collide. After they collide and you see the vectors change, click "pause." Click "rewind" and watch the momenta box during the collision. Watch it more than once if needed by using "play", "rewind", and "pause". Zoom in on the vectors in the momenta box with the control on the right of the box to make it easier to see if necessary.
   - What happens to the momentum of the red ball after the collision?
   - What about the green ball?
   - What about the total momentum of both the red and green ball?

4. Change the mass of the red ball to match that of the green ball.
   - Which ball has greater momentum now?
   - How has the total momentum changed?
   - Predict what will happen to the motion of the balls after they collide.

5. Watch the simulation, and then pause it once the vectors have changed.
   - What happens to the momentum of the red ball after the collision?
   - What about the green ball?
   - What about the total momentum of both the red and green ball?

6. Now change the elasticity to 100%. Predict the motion of the balls after the collision.

7. Watch the simulation, and then pause it once the vectors have changed.
   - What happens to the momentum of the red ball after the collision?
   - What about the green ball?
   - What about the total momentum of both the red and green ball?

8. Make up your own collision scenario (you may use the two-dimension setting) and make predictions about the movement of the balls. Diagram and describe it.

9. Experiment by running additional simulations. Record the data for at least three (3) additional simulations (each extra simulation = 5 points, maximum of 5 trials).

Data
Sample data table

Trial	Mass of Red Ball	Mass of Green Ball	% Elasticity	Red and Green Momentum Vectors before Crash	Red and Green Momentum Vectors after Crash	Change in Total Momentum during Simulation? (yes or no)
1						

Conclusion

Organize and answer all the questions posed in the Procedure section.

12. Car Crash Inquiry Lab

Introduction
A traffic accident occurred in a 35 km/hr speed limit zone on Cook Street in which a 3000 kg Cadillac Escalade SUV rear-ended a 2000 kg Subaru Outback Wagon stopped at a stop sign. The entire police investigative division went on vacation to Bora Bora to relax, so the Chief of Police has contracted with you and your team of experts to determine what happened and what traffic laws were broken.

Task
Your team will provide the Chief of Police with a detailed accident report (including all equations and work) that includes mass, velocity, and momentum of both vehicles both prior to and after the collision. Further, the Chief of Police requested you create a visual demonstration/re-creation of what happened to assist with the insurance company's investigation. This may take the form of a PowerPoint presentation, a 2'x 3' poster, a website, or an annotated digital video.

The Chief of Police provided the following diagram drawn by police officers at the accident site:

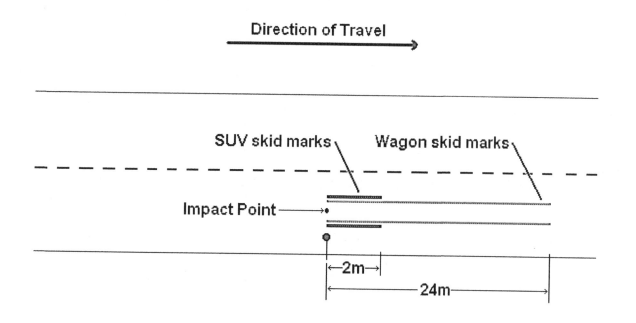

Procedure

1. Each group member will choose one of the following roles. The group's success will depend upon how well each individual accomplishes their responsibilities.
   - *Auto Expert:* This individual will research the physics of linear motion and determine, based upon accident site analysis, how fast the SUV and wagon were moving immediately following the collision.
   - *Collision Expert:* This individual will research the physics of basic collisions and determine, based upon data provided by the auto expert, how fast the SUV was initially moving.
   - *Investigator:* This individual will research elastic and inelastic collisions, then analyze the accident scene as well as reports from the auto expert and collision expert in order to determine what type of collision occurred.

2. Fulfill your individual roles. Before you can put together a comprehensive picture of what happened, each individual member of the group must visit several websites and research their portion of the accident to understand what is required and what steps you must take to accomplish your mission. Click on the links below to guide you through your role-specific tasks.

3. Design your written accident report. This should include a one-paragraph introduction, a written step-by-step summary of what you believe occurred (including any appropriate diagrams), any assumptions you made during the analysis, and a listing of what traffic laws were broken, and by whom. Include a final paragraph describing why you would or would not be interested in a career in accident investigation.

4. Design your final presentation, which visually and verbally explains to a non-technical audience in a professional manner what your group believes occurred at each step of the accident.

Evaluation
Work will be evaluated based on the attached rubric. Each group will be assessed based on criteria in the grading rubric for a total of 20 possible points.

Conclusion
Once the project is completed, students should have a solid understanding of momentum, conservation of momentum, and inelastic and elastic collisions. Further, students will have connected previous kinematics work with the newer topics of collisions and energy. Finally, students will have been introduced to careers in accident investigation.

Auto Expert Responsibilities

As the team expert in automobile kinematics and performance, it is your job to analyze the accident site and determine the velocities of both the SUV and the wagon immediately following the collision.

1. Review the kinematic equations for one-dimensional motion. Link to:
   - A Plus Physics 1D Kinematics Page
     http://www.aplusphysics.com/courses/regents/kinematics/regents_kinematic_equations.html
   - Physics Classroom Kinematic Equations
     http://www.physicsclassroom.com/Class/1DKin/U1L6a.cfm
   - Physics Lab Online Kinematics
     http://dev.physicslab.org/Chapter.aspx?cid=21

2. Analyze the diagram of the collision scene below. Note that the acceleration of a Subaru Outback Wagon with the brakes locked is -3 m/s², and the acceleration of a Cadillac Escalade SUV with the brakes locked is -2 m/s².

3. Using the information provided, and looking at just the time period **AFTER** the collision, find the initial velocities of both the SUV and the wagon prior to their deceleration to rest. Give this information to your team members.

4. Create a one-page type-written report (including appropriate diagrams and equations) explaining your work and showing your derivations.

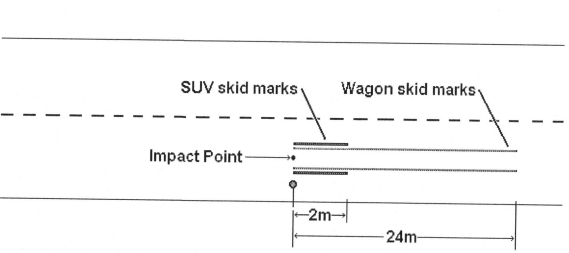

Collision Expert Responsibilities

As the team expert in collisions, it is your job to research the principles of conservation of momentum. You will use this knowledge, along with data provided by the auto expert, to determine the initial velocity of the SUV prior to the collision.

1. Research the qualitative and quantitative definitions of momentum, as well as the law of conservation of momentum. Link to:
   - A Plus Physics Momentum Page

http://www.aplusphysics.com/courses/regents/momentum/regents-conservation-momentum.html
- Physics Classroom Momentum Page
  http://www.physicsclassroom.com/class/momentum/u4l1a.cfm
- Physics Lab Momentum Page
  http://dev.physicslab.org/Document.aspx?doctype=3&filename=Momentum_Momentum.xml

2. Using the information provided by the auto expert, calculate the total momentum of the wagon/SUV system AFTER the collision. How does this compare to the total momentum of the system BEFORE the collision?

3. Knowing the total momentum of the system BEFORE the collision, calculate the initial velocity of the SUV (Hint: What is the initial velocity of the wagon? What is its initial momentum?)

4. Create a one-page type-written report (including appropriate diagrams and equations) explaining your work and showing your derivations.

Investigator Responsibilities

As the team coordinator and head investigator, it is your job to research elastic and inelastic collisions and, using reports from the auto expert and collision expert, determine whether this collision was elastic or inelastic in nature. Further, you are responsible for coordinating the work of the entire team to insure each team member is providing the appropriate information to put together a comprehensive picture of what happened during the accident.

1. Research the qualitative and quantitative definitions of kinetic energy, as well as what differences exist between elastic and inelastic collisions both qualitatively and quantitatively. Link to:
   - A Plus Physics Kinetic Energy
     http://www.aplusphysics.com/courses/regents/kinematics/regents_kinematics.htm
   - A Plus Physics Collisions
     http://www.aplusphysics.com/courses/regents/momentum/regents-collisions.html
   - Physics Classroom Kinetic Energy
     http://www.physicsclassroom.com/Class/energy/u5l1c.cfm
   - Physics Lab Kinetic Energy
     http://dev.physicslab.org/Document.aspx?doctype=3&filename=WorkEnergy_MechanicalEnergy.xml

2. Using the information provided by the auto expert and collision expert, determine whether the accident you are investigating is elastic or inelastic. Examples of similar situations may be used for reference:
   - Physics Classroom Collisions
     http://www.physicsclassroom.com/Class/momentum/u4l2d.cfm

- Elastic and Inelastic Collisions
  http://www.walter-fendt.de/ph14e/collision.htm

3. Create a one-page type-written report (including appropriate diagrams and equations) explaining your work and showing your derivations.

Car Crash Grading Rubric

Group Names: _____

Position	Beginning 1	Developing 2	Very Good 3	Exemplary 4	Score
Auto Expert's Report	Given information correctly identified, kinematic equation chosen appropriately.	Kinematic equation applied effectively, correct velocities identified.	Correct velocities identified, report explains work performed.	Correct velocities identified, comprehensive professional report including equations and diagrams.	
Collision Expert's Report	Given information correctly identified, conservation of momentum utilized.	Conservation of momentum applied effectively, correct velocities identified.	Correct velocities identified, report explains work performed.	Correct velocities identified, comprehensive professional report including equations and diagrams.	
Investigator's Report	Given information correctly identified, kinetic energy formula utilized.	KE definition applied effectively, correct type of collision identified.	Correct type of collision identified, report explains work performed.	Correct type of collision identified, comprehensive professional report including equations and diagrams.	
Group's Accident Report	Report is missing key sections or information.	Report includes all required sections.	Report includes all required sections with appropriate diagrams and equations.	Professional comprehensive report with accurate answers, including appropriate diagrams and equations.	
Final Presentation	Final presentation does not accurately represent accident.	Final presentation accurately represents accident.	Final presentation accurately represents accident, with appropriate justification and explanation.	Final presentation accurately represents accident, with appropriate justification and explanation. Assumptions and areas for improvement critically explained.	

TOTAL points _____

_____ **Total points x 5 = _____ %**

13. Conservation of Energy Lab

Purpose
To use an incline plane and golf ball to investigate the conservation of mechanical energy

Equipment
Incline plane, golf ball, stopwatch, meter stick

Procedure
1. Construct an incline plane with your track.

2. Find the mass of a golf ball.

3. Release the golf ball from the top of the inclined plane, timing when it reaches the bottom.

4. Repeat step 3 at least two more times until you have three trials with consistent measurements.

Data
Record your data in a table.

Calculations
For each of the following calculations, show your work.

1. Calculate the velocity (m/s) of golf ball when it's at the bottom of the ramp.

2. Calculate the potential energy of the golf ball at the top of the ramp.

3. Calculate the kinetic energy of the golf ball at the bottom of the ramp.

Conclusions
1. If mechanical energy was conserved as the golf ball went down the ramp, how should the amount of potential energy at the top of the ramp compare to the amount of kinetic energy at the bottom of the ramp?

2. Find the percent of difference between your calculated values for potential and kinetic energy. Was mechanical energy conserved?

3. Explain how your lab results are consistent with the law of conservation of energy. Where did the missing energy go?

14. Energy to Work Lab

Purpose
- To understand work and its relation to energy
- To understand how energy can be transformed from one form into another
- To compute the power from the rate at which work is done

Equipment
Balloon, whiffle ball, meter stick, stop watch

Introduction
The universe is made of matter and energy. Matter is substance and energy moves the substance. The concept of matter is easy to comprehend, because we can see, touch, smell, or measure it. Energy, on the other hand is an abstract concept. We cannot see, touch, or smell it, but we can measure it. There are various units used to measure energy. A non-metric unit of calorie (cal) is one of them. The energy equal to 1 *calorie* is defined as the amount of heat required to raise the temperature of one gram of water from 14.5°C to 15.5°C. When we work with nutritional values and food, a unit of **Calories** (Cal - capitol C) is used. Relation between cal and Cal is: 1 Cal = 1000 cal.

Metric unit for energy is Joule (J). The energy of 1 Joule is equal to amount of heat needed to raise the temperature of 1 g of water 1 °C. The relation between the three units is:

$$1 \text{ Cal} = 1000 \text{ cal}$$
$$1 \text{ cal} = 4.19 \text{ J}$$
$$1 \text{ Cal} = 4190 \text{ J}$$

In the physical world, the possession of *energy* by an object means it has an ability to do *work*. Work done is a measure of the "effect" the application of a force produces. If the applied force and the displacement of the object are in the same direction, then the work done is given by, Work Done = Force x Distance.

Mechanical energy has several different forms. *Elastic Potential Energy* is the stored energy by virtue of an object's *configuration*. When you stretch a spring, you are doing work on the spring and in turn the spring stores that work in the form of elastic potential energy.

Gravitational Potential Energy, on the other hand, is the stored energy by virtue of an object's height (position). When the gravitational force is the only force on an object, the gravitational potential energy is calculated from:
Gravitational Potential Energy = Weight x Height.

Energy that exists by virtue of an object's motion is called the *Kinetic Energy*. The *law of conservation of energy* is a universal principle that says the total energy of a system always remains constant. In other words, energy cannot be created or destroyed but it can be converted from one form into another.

When work is done on an object, any of the following can happen:
- The object may, in turn, do work on another object,
- The object's speed may increase (gain in kinetic energy),
- The object's temperature may rise (gain in thermal energy),
- The object may store the energy for later use (gain in potential energy),
- The object may rise in the earth's gravitational field (gain in gravitational potential energy).

In many situations not only is the amount of work done important, but also how slowly or how quickly it is done is also important. The rate at which work is done or energy is transformed is called *Power*. Therefore, the power is calculated as follows:

Power = Work Done/Time = Energy Used/Time.

In the first part of this lab you will learn how the work done on an object is stored as potential energy of that object. In the second part you will figure out the work done during various activities and compute the power expended.

Procedure
Part I
Answer the questions given in the Lab Experiment Questions and Data section after doing the following:
1. Blow up a balloon and then release it.
2. From the ground, raise the ball to a height of about two (2) meters and release it.

Part II
1. Go to the nearest stairway. Measure the height of one step and the number of steps to the 2nd floor. Record your measurements.
2. Have one of your partners climb the stairs slowly. Record the time taken.
3. Next, climb at a faster speed and record the time.

Lab Experiment Questions and Data
Part I
Answer the following questions.
Energy storage in a balloon:
1. Did you do work when you blew up a balloon? How do you know?
2. In what form of energy is stored in a blown up balloon?
3. How can you get the stored energy out of the balloon? Where does it go?

Energy storage in a ball:
1. Did you do work when you raised the ball to two (2) meters height? How do you know?
2. In what form is the energy stored at two (2) meters height?
3. What happens to this stored energy as the ball falls to the ground? Explain.

Part II
Use the following to estimate your weight in newtons:
        100lb (or less) use 400N
        125lb use 556N
        150lb use 667N
        175lb use 778N
        200lb use 889N
        225lb use 1001N

- Your weight _____ N (newtons)

- Height of one step _____ m (meters)

- Number of steps to 2nd floor _____

- Time taken while climbing slowly _____(seconds)

- Time taken while climbing faster _____(seconds)

Calculations
Write the formula/equation, then solve showing calculations. Calculate the following:
- Total distance to 2nd floor
- Work done in climbing to 2nd floor
- Power output (slow)
- Power output (fast)

Conclusion
Answer the Lab Experiment Questions and the following:

1. When you drop the whiffle ball from a certain height, does it return to the same height after bouncing from the floor? Does that violate the conservation of energy principle? Explain.

2. In Part II, is the work done greater, smaller or the same when you climb fast or slow? What about the power output? Explain.

3. Given that 4190 Joules = 1 Food Calorie, calculate the number of Calories you used by climbing to the 2nd floor. How many stories (approximately) would you need to climb to turn 100 Calories (energy equivalent of 1 cookie!) of energy into gravitational potential energy? Comment on the results.

15. Torque Lab

Purpose
To explore the concepts of torque and to calculate balancing forces

Equipment
Ring stand, support arm, string, spring scales, weights, meter stick

Procedure
1. Record all data in your lab journal. See the "Suggested Format..." for collecting data at the end of this lab.

2. Calibrate your spring scale.

3. Use a spring scale to weigh your meter stick. Do this by hanging the meter stick from a string and suspending the string from the scale.

4. Set up a ring stand and support arm. Hang the meter stick from the support arm using a string. Make a note of where the string is located on the meter stick when it is balanced. This is the center of gravity reading, you will need this later. **Show the location of your fulcrum in each of the diagrams.**

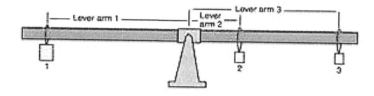

Working Backwards with Two Weights

1. Take two weights, they can be the same or different. Use strings, if needed, to hang both of the weights and arrange the weights so the meter stick is balanced. Draw the arrangement with weights in your lab journal. The fulcrum is shown as a triangle below the scale.

2. Work backwards to show the torque from each of the weights causes a net torque of zero. You will need to use the distance from the fulcrum (string holding the meter stick) to each of the weights. Show all of your work.

Working Backwards with Three Weights
1. Choose three weights and repeat the same experiment. This may take a bit of trial and error. Think and be patient. Draw the arrangement with weights and readings in your lab journal.

2. Work backwards to show the torque from all three of the weights causes a net torque of zero. Show all of your work.

Working Forward with Two Weights

1. Keep the meter stick support string at the center of gravity. Choose two of the weights, pick one weight and choose a location for that weight. Draw the arrangement with weights and readings in your lab journal.

2. Knowing the net torque on the meter stick is zero, calculate the location of the second weight.

3. Place the weight at the calculated location. Is your system balanced? In a different color, show where you needed to move the weights to balance the system.

Working Forward with Three Weights

1. Keep the meter stick support string at the center of gravity. Choose the lighter two of the three weights and place them on one side of the meter stick balance. Draw the arrangement with weights and readings in your lab journal.

2. Knowing the net torque on the meter stick is zero, calculate the location of the third weight.

3. Place the weight at the calculated location. Is your system balanced? In a different color, show where you needed to move the weights to balance the system.

Unbalanced Meter Stick 1

1. Move the fulcrum (the string holding up the meter stick) to the 25 cm point. You now have a weight at the center of mass of the meter stick equal to the weight of the meter stick. Draw your starting point.

2. Using a single weight of your choice, calculate where to place the weight so the system is balanced. Show all your work.

3. Place the weight at the calculated location. Is your system balanced? In a different color, show where you needed to move the weights to balance the system.

Unbalanced Meter Stick 2

1. Repeat the above experiment using two or three weights and an unbalanced meter stick. Move the fulcrum to any other point except the center of mass. Draw your starting point.

2. Using any two or three weights of your choice, calculate where to place the weight so that the system is balanced. Show all your work.

3. Place the weight at the calculated location. Is your system balanced? In a different color, show where you needed to move the weights to balance the system.

Suggested Format for Torque Lab Data

Working Backwards with Two Weights

Working Backwards with Three Weights

Working Forward with Two Weights

Working Forward with Three Weights

Unbalanced Meter Stick

Unbalanced Meter Stick 2

16. Rolling Cylinders & Spheres Lab

Purpose
To explore the difference in acceleration for different rolling objects, to determine what factors affect the acceleration of an object rolling down an incline plane, and to compare the rolling speeds of several cylindrical and spherical objects down an incline plane

Equipment
Objects of a variety of shapes and sizes, inclined plane, stopwatch, meter stick, protractor

Theoretical Values
The best way to look at the motion of rolling objects is from an energy viewpoint. The work-energy theorem states: $W = \Delta E = \Delta K_{trans} + \Delta K_{rot} + \Delta U$

Where work is: $W = F_{net} \Delta r \, cos\theta$

Notice that in this expression, there are two types of kinetic energy, translational and rotational. Translational K is the energy associated with the motion of the center of mass of the object. Rotational K is associated with motion about the center of mass.

$$K_{trans} = \frac{1}{2} mv^2$$

$$K_{rot} = \frac{1}{2} I\omega^2$$

Where I is the "moment of inertia," the rotational equivalent of mass and it depends on the shape.

For a hollow cylinder (a hoop): $I_{hoop} = MR^2$
For a solid cylinder: $I_{cylinder} = \frac{1}{2} MR^2$
For a solid sphere: $I_{sphere} = \frac{2}{5} MR^2$
For a spherical shell: $I_{sperical \, shell} = \frac{2}{3} MR^2$

Now, if it is rolling, then the following is true: $v = \omega R$. Using this, you can combine the two kinetic energies. If the object is rolling down the incline plane, you could say there is no work done on the object. You can use the change in gravitational potential as: $\Delta U = mg\Delta y$.

From all this, you could get the speed at the bottom of an incline. Assuming the acceleration is constant, you could find this using the initial speed and final speed. Now you can compare it to your data.

Procedure

1. Pick two objects with the same shape but different radii. Place them side-by-side on the incline and release them from rest at the same time. How does the speed depend on the radius?

2. Pick two objects with the same shape but different masses. Place them side-by-side on the incline plane and release them from rest at the same time. How does the speed depend on the mass?

3. Now compare the rolling speeds of objects of different shapes (cylindrical shell, solid cylinder, spherical shell, solid sphere). What are your results?

4. Are your experimental observations consistent with the general equation for the speed you derived above? Explain.

5. How would you design a rolling object that was faster than any of the ones you used here?

Conclusion

1. A solid cylinder of radius $R = 2$ cm and mass $M = 100$ g rolls without slipping down an inclined plane. If it starts from rest at an elevation $h = 25$ cm, what is its translational speed when it reaches the base of the incline? Repeat the calculation for a solid cylinder with $R = 4$ cm and $M = 100$ g. Repeat again for $R = 2$ cm and $M = 200$ g. Hint: Use conservation of energy and the fact that $\omega = v/R$.

2. Assume the rotational inertia of a round object can be written as $I = kMR^2$, where k is a shape factor. Derive a general expression for the speed of the object in terms of k and the initial height, h.

17. Centripetal Force Lab

Purpose
To investigate the relationship between the speed of an object in uniform circular motion and the centripetal force on the object

Equipment
Centripetal force apparatus, washers or masses, meter stick, stopwatch, whiffle ball

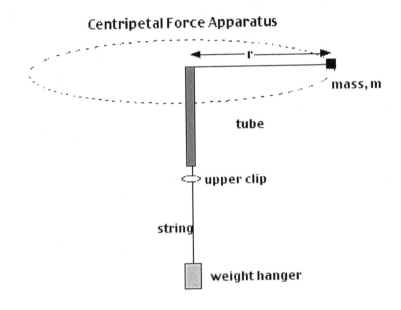

Centripetal Force Apparatus

Discussion
A force, which pulls an object toward the center of a circle, is called *centripetal force*. How much centripetal force needs to be exerted to cause an object to move in a circle? Your experience should tell you the amount of centripetal force you need to exert depends on:

- the mass of the object you are whirling - heavier objects require more force,
- how fast you are whirling it - going faster requires more force, and
- the radius of the circle.

The formula states: $F_{cent} = \dfrac{mv^2}{r} = mr\omega^2$

but how can you verify the quantitative relationship between centripetal force and mass, speed, and radius? This question cannot be answered all at once, since a scientific experiment is designed to vary one quantity (holding all others constant) and measure its effect on **one** other quantity. The most difficult quantity of mass, speed, and radius to hold constant from trial to trial in this experiment is the speed of the object. Therefore, it is easiest to study the effect of speed on centripetal force, since it is relatively easy to hold the mass of the object and the radius of the circle constant.

Using an apparatus similar to the one pictured, you will measure the effect of speed on centripetal force. You can hold the mass constant during a set of trials by always whirling the same object. You can keep the radius of the circle constant (with a little practice) by keeping the upper clip a fixed distance below the tube while whirling the object.

Procedure

1. Place a small number of masses or washers on the bottom clip of the apparatus. This part of the apparatus hangs straight down, and the weight of the washers supplies the centripetal force.

2. Practices whirling the ball until you can keep the upper clip a short distance below the bottom of the tube while the ball whirls. **IMPORTANT! If the clip touches the bottom of the tube, the masses are no longer supplying the centripetal force!** If the clip rises or falls appreciably as the ball whirls, the radius of the circle is changing. Practice!

3. Use a stopwatch to measure the time taken for a reasonable number of revolutions (20 to 30). Record all necessary data in a table.

4. Change the masses/number of washers on the bottom clip (centripetal force) and repeat steps 3 and 4. Repeat for several different weights. Record the data.

5. Change the position of the upper clip to change the radius of the circle. Repeat the experiment for this radius. Be sure to indicate where the radius changes in your data table.

Calculations and Graphs

1. Calculate the period of revolution, T (the time to go around once) for each trial. Show a sample calculation.

2. Calculate the linear speed, v, of the ball for each trial. Show a sample calculation.

$$(\text{Note: } v = \frac{2 \pi r}{t})$$

3. Theoretically, the centripetal force should be directly proportional to the **square** of the speed. To check this, add a column to your data table for v^2.

4. Construct a graph of centripetal force versus v^2. Remember it is customary to put the quantity changed (force, in this case) on the horizontal axis, and the quantity that changes by itself (speed) on the vertical axis. Be sure you pick the largest convenient scale for the graph and draw the **best smooth curve** through the data points.

Conclusion

1. Is the graph of centripetal force versus speed squared a straight line?

2. What can you say about the relationship between centripetal force and speed?

18. Coulomb's Law Lab

Purpose

To investigate electrostatic charges and use Coulomb's Law and Newton's Second Law to calculate the number of electrons on the surface of a charged object

Equipment

Balloons, wool cloth, black rod, electroscope, string, meter stick

Information

The electron and proton charges and masses are provided below.

Electron (e)	Proton (p)
$m_e = 9.1 \times 10^{-31}$ kg	$m_p = 1.67 \times 10^{-27}$ kg
$e = -1.602 \times 10^{-19}$ C	$e+ = 1.602 \times 10^{-19}$ C

Procedure

The black rod and wool cloth will leave a negatively charged rod and the wool cloth will be positively charged. Charge the electroscope using the black rod. This will give a negatively charged scope.

Weigh the balloons, using the triple beam balance. Balloon mass is _____ kg.

Blow up two balloons tie a knot and charge them by rubbing with the wool cloth. Using your electroscope discover whether the balloon is positively or negatively charged. Balloon charge is _____ (+/-).

Tie the balloons to the two strings hanging down from the ceiling. They will most likely discharge during this process, so once they are hung, charge them again. The balloons should separate from each other and make a 10° to 20° degree angle at the ceiling.

N.B. Air conditioning vents may effect this procedure. It may be done underneath a desk or lab table to minimize this problem.

Measure the length from the ceiling to the center between the balloons.

Measure the angle by measuring the mid-point to mid-point separation of the balloons. You should be able to get this to an accuracy of 1 or 2 cm. Use half of this separation (d/2) and the length (L) to determine the angle.

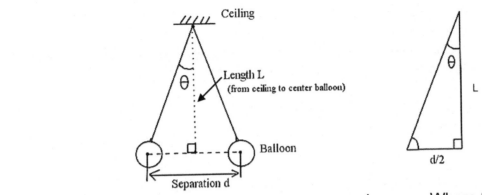

L = _____ m   d = _____m   θ = _____degrees. Where tan θ = d/(2L).

Start with a free body diagram (FBD) for one balloon. Treat the balloon as a point (which you can think of as the center of the balloon).

Calculations

Below is a free body diagram of the forces on the balloon. There is tension and electrostatic repulsion. These forces must balance for the balloon to be stationary. Note: $k_e = 1/4\pi\varepsilon_o$, therefore $\varepsilon_o = 1/4\pi k_e$, where ε_o is the permittivity of free space.

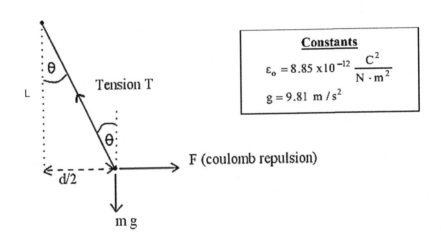

Now consider balancing horizontal and vertical forces.

1) Horizontally: $T \sin \theta = F$

2) Vertically: $T \cos \theta = mg$

3) Dividing (1) by (2): $\tan \theta = \dfrac{F}{mg} = \dfrac{d}{2L}$

4) Coulomb's Law says: $F = \dfrac{q^2}{4\pi\varepsilon_0 d^2}$

5) This implies that,
$$d^3 = \frac{Lq^2}{2\pi\varepsilon_0 mg}$$

6) Hence:
$$q = \sqrt{\frac{2\pi\varepsilon_0 mgd^3}{L}}$$

Show your work to derive equation (6) using equations (3) and (4).

Find q, then divide by the electronic charge "e" to find the number of electrons either extra on the balloon or missing from the balloon, depending on whether it is +/- . The charge is a very small number. It depends on the humidity. A dry day is best.

Number of electrons, N = _____. This should be a very large number.

Conclusion

1. Which force was NOT on the free body diagram?
   a. Tension
   b. Weight
   c. Coulomb force
   d. Normal force

2. We rub completely around the balloons to give them charge. Could achieve an equal charge distribution by just rubbing one side of the balloon?
   a. Yes, because the charge will spread out equally around the balloon
   b. No, because the balloon is an insulator and the charge will stay local.
   c. Yes, because the charge can move through the air inside the balloon to the other side.
   d. No, because the protons removed from the balloon to the wool cloth do not shift.

3. You measured the radius between the balloons 3 times and found values of 13.0 cm, 13.5 cm, and 15.0 cm. The correct way to present the average value with error would be;
   a. 13.8 ± 2.0 cm
   b. 13.8 ± 1.0 cm
   c. 14.0 ± 1.0 cm
   d. 14.0 ± 2.0 cm
   e. 13.5 ± 1.0 cm

4. The picture below shows a situation where balloon B has twice the angle of balloon.

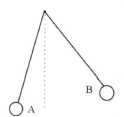

This could occur if:
a. Balloon A weighs more than balloon B, both same charge.
b. Balloon B weighs more than balloon A, both same charge.
c. Balloon B has more charge than balloon A both weigh the same.
d. More than one answer above is correct.
e. This cannot happen

5. Two small charged objects attract each other with a force F when separated by a distance d. If the charge on each object is reduced to one-fourth of its original value and the distance between them is reduced to d/2 the force becomes:
   a. F/16
   b. F/8
   c. F/4
   d. F/2
   e. F

6. The units of $\frac{1}{4\pi\varepsilon_0}$ are:

   a. $N^2 C^2$

   b. $\frac{N \cdot m}{C}$

   c. $\frac{N^2 \cdot M^2}{C^2}$

   d. $\frac{N \cdot m^2}{C^2}$

   e. $\frac{m^2}{C^2}$

7. The leaves of a positively charged electroscope diverge more when an object is brought near the knob of the electroscope. The object must be:
   a. a conductor
   b. an insulator
   c. positively charged
   d. negatively charged
   e. uncharged

8. Two particles, each with charge Q, and a third particle, with charge q, are placed at the vertices of an equilateral triangle as shown. The force on the particle with charge q is:
   a. Parallel to the left side of the triangle
   b. Parallel to the right side of the triangle
   c. Parallel to the bottom side of the triangle
   d. Perpendicular to the bottom side of the triangle
   e. Perpendicular to the left side of the triangle

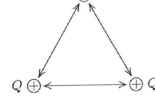

19. Direct Current Electrical Circuits Lab

Purpose
To use circuit symbols, construct series and parallel circuits, and use voltmeters and ammeters correctly in circuits

Equipment
Bulbs, voltage sources, voltmeter, ammeter, wires, switch

Procedures
1. Sketch a schematic series circuit diagram with two (2) bulbs, 3V source, switch, and ammeter and voltmeter properly connected.

2. Wire the series circuit. Using the ammeter and voltmeter, record the voltage and current. Complete a VIRP table for this circuit.

3. Repeat steps 1 and 2 for three (3) bulbs in series.

4. Unscrew one (1) bulb at a time in the 3-bulb circuit and record the effect.

5. Sketch a schematic parallel circuit diagram with two (2) bulbs, 3V source, switch and ammeter and voltmeter properly arranged to read total current and voltage across a single bulb. Assume all bulbs are similar.

6. Wire the parallel circuit. Using the ammeter and voltmeter, record total current and potential difference for *each* bulb on a VIRP table. Assume the current is equally divided.

7. Repeat steps 1 and 2 for three (3) bulbs in a parallel circuit.

8. Unscrew one (1) bulb, then two (2) bulbs in your 3-bulb circuit. Record and explain the results.

Data
Construct data tables and record data.

Calculations
1. For each circuit, calculate the total energy (W) used if each circuit operated for 60 minutes.

2. For each circuit, calculate the total charge (q) that would pass in 60 minutes.

3. Write a paragraph comparing current, voltage, power, and resistance for your two VIRP tables.

Other Information Required

In your lab report, include the following:

1. Write the rules for series circuits.

2. Explain the correct connections for ammeters and voltmeters.

3. Write the rules for parallel circuits.

4. What happens to the total resistance, amount of current, and power in a parallel circuit as more and more resistors are added? What happened to the brightness of each bulb during the lab?

5. State the differences between the 3-bulb series and 3-bulb parallel circuits with regard to I, V, R, P, and energy.

20. Resonance Lab

Purpose
To use resonance to make pendulums swing without any initial push

Key Concept
Resonance is the back-and-forth motion that becomes especially strong when a pendulum is repeatedly pushed at its natural frequency. This effect can happen with any system with a natural vibrating frequency (e.g., earthquake toppling a building, opera singer breaking a crystal glass, etc...).

Equipment
Two rulers, table, meter stick, several meters of string, five(5) small weights (e.g., washers), masking tape, protractor, stop watch, computer with Internet access for generating "beats" with sound waves

Introduction
Any object with a natural frequency for moving back and forth can experience *resonance* if it is pushed by a force that matches that frequency. When resonance occurs, the amplitude of the object's motion can become quite large. Without resonance, even a big force will not do much to keep the object moving.

Imagine you push a child on a swing. The swing acts like a pendulum and has a natural frequency for swinging back and forth, depending on how long it is. How would you push the swing to make it go higher and higher? You do not just pull it up once and let go. Instead, you keep pushing every time the swing comes back towards you. In fact, you have to push with the exact same frequency as the swing – for every swing back and forth you make one pushing motion. If you were to push with a different, random frequency, you would not have much luck in getting the swing to go high.

In general, an object with a natural frequency will move with much larger amplitude if it is pushed by a force with the same frequency. This applies not only to pendulums but to any regular motion. For instance, the crystal in a goblet has a natural frequency of vibration, and a good opera singer can sing a note of exactly the right frequency to make it start vibrating and shatter. When designing a suspension bridge engineers take care its natural swinging frequencies do not match any frequency it is likely to encounter in a windstorm. In this lab, you will experiment with resonance in pendulums.

Haunted Pendulums: Make a pendulum move without touching it!

1. Create pendulums with one washer. Attach it to the string across the rulers, on the other sides of the long pendulum (see diagram). Carefully adjust the length of your new pendulums to be as close as possible to the length of each other.

2. Pull back one of the short pendulums and let it swing.
   - *What happens to the other short pendulum?*
   - *What happens to the central long pendulum?*
   - *Since you did not touch it, where does the force that makes the other short pendulum swing come from?*
   - *In which of the two pendulums you did not touch does resonance occur? Why did it happen for that one?*
   - *If you have two pendulums of the same length but with a different number of washers at the end, do you expect to see resonance?*

3. Now move the central long pendulum out of the way by draping it over a ruler or the table.
   Repeat the experiment by pulling back one of the short pendulums, releasing it.
   - *If you wait a little longer, what happens to the pendulum you pulled back?*
   - *Make a rough plot of how the original pendulum you pulled swings over time.*
   - *Angle 0 corresponds to the pendulum hanging straight down. Positive angles are with the pendulum pointing towards you; negative angles are with the pendulum pointing away from you. Time 0 is when you first let go of the pendulum.*

This phenomenon is called **beats**. It happens when you have two different frequencies contributing to a back-and-forth motion. You can hear it with sound if you play two violin strings that are very close to the same frequency but not quite – the sound will seem to waver between loud and soft, in the same way that these pendulums alternated between large amplitudes and small ones. This phenomenon occurs because when two waves have different frequencies, they interfere with each other.

Sometimes, the waves line up so the interference is said to be **constructive** (making the net wave bigger than either of the individual waves); other times, the waves oppose each other so the interference is said to be **destructive** (making the net wave smaller than either of the individual waves).

4. A demonstration of sound beats can be done using an online "tuning fork" that will play a perfect pitch: http://www.tictone.com/online-tuning-fork.php Open up two browser windows such that both are playing the "A4" pitch at 440 Hz. Adjust one of the pitches a couple Hz (e.g. to 442 Hz). You will hear beats. *If you increase the difference between the pitches of the two simultaneous notes, how does this affect the frequency of the beats?*

Pendulum Mind-Control: Control which pendulums swing!

5. Now attach a "handle" string at both ends to either side of the string between the rulers. Your pendulum setup will now look as shown below.

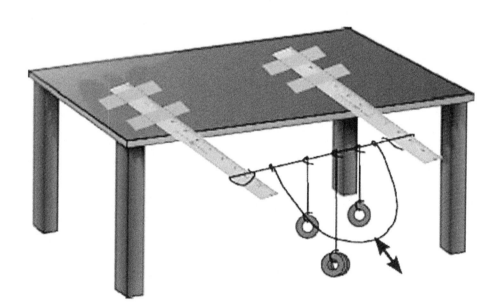

6. Tug rhythmically on the center of your handle-string.
   - *Can you make the two short pendulums swing high without the long pendulum swinging much?*
   - *Can you make the long pendulum swing high without the short ones swinging much? Can you make only one of the short pendulums swing without the other one swinging?*
   - *What do you have to change in order to control which pendulums are swinging?*

Conclusion

1. Old-fashioned grandfather clocks use a long pendulum that swings back and forth to keep time. When the metal of the pendulum is heated slightly, it stretches so the pendulum becomes longer. On a warm summer day, is a grandfather clock likely to run fast, run slow, or keep good time?

2. Imagine a giraffe, a mouse, and an ant walking side-by-side. For every step the giraffe takes, the mouse will take many more, and an ant will take even more. In general, why do animals with short legs tend to move them more frequently than animals with long legs?

3. Soldiers are ordered to break step when going across a bridge. Why is this? Why are the bridges unlikely to collapse from crowds of ordinary people walking across?

4. When designing a tall building in a region where earthquakes can happen, engineers have to take into account the typical frequencies associated with the back-and-forth vibration of the ground in an earthquake. Why?

21. Exoplanets and Habitable Worlds Lab

Introduction
Until the discovery of the first extrasolar planet (or *exoplanet*) in 1994, we knew almost nothing about the existence of other solar systems. Now the search to find *habitable*, Earth-like planets that can support life will be one of the accomplishments of this century. This exercise will introduce you one of the best techniques for finding such worlds, which was employed by the European Space Agency's *COROT* mission in 2007 and NASA's *Kepler* mission launched in 2008.

Purpose
To complete an online exercise to understand how astronomers are searching for habitable Earth-like planets

Planet-Hunting among the Stars
Even the nearest solar systems are too distant for conventional telescopes to separate the bright, central stars from any planets orbiting them. Most methods of exoplanet detection use the effect of a planet's gravity on its star, which causes the star to "wobble" as the planet swings around it. The wobble may be detected by measuring the star's position relative to others, or (more commonly) by using the Doppler effect to measure the small back-and-forth change in the star's velocity. Both of these methods are effective for finding high mass "Jupiters," but are poor at detecting smaller, Earth-mass planets, which have only a tiny effect on their central star's motion.

The Transit Method
The transit method looks for a drop in the brightness of a star when a planet passes in front of it, an event astronomers call a *transit*. This method will not find every planet, only those that happen to cross our line of sight from Earth to the star. But with enough sensitivity, the transit method is the best way to detect small, Earth-size planets and has the advantage of giving us both the planet's size (from the fraction of starlight blocked), as well as its orbit (from the period between transits). Since a transit only lasts a few hours, continuous monitoring is required. A triumph of the transit method occurred in 1999 when the light curve (a plot of brightness vs. time) of the star HD209458 showed a large exoplanet in transit across its face. The star's brightness as seen from Earth drops by a few percent every time the orbiting exoplanet crosses in front of it, as shown. As the planet leaves the disk of the star, the brightness increases again. The time interval between such dips in brightness tells us the planet's orbital period around its star. Using the mass of the star and Kepler's laws of planetary motion, we can then find the size of its orbit, i.e. its average distance from the star.

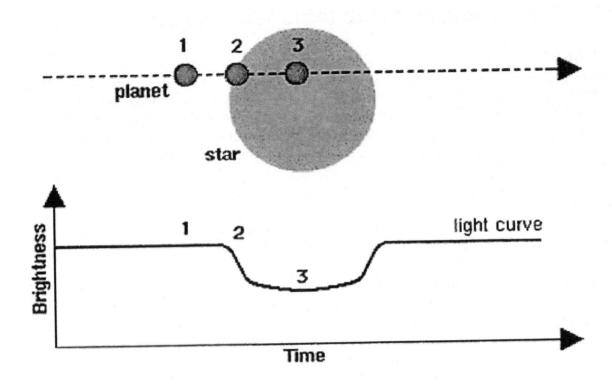

The Habitable Zone

For life to develop on a planet, it must be "not too hot, not too cold, but just right" for liquid water to exist on its surface. The range of distances from the star for which this is possible is known as the Habitable Zone. As expected, if the star is massive and luminous, this zone is farther away compared to that for a low-mass, cool star. Since the transit method allows us to calculate the size of the planet's orbit, we can immediately tell if this newly discovered world lies in the habitable zone, making it a candidate for life to develop.

We must await results from the *COROT* or *Kepler* missions for evidence that Earth-size planets exist around other stars. In this online laboratory exercise, we use *simulated* light curves of stars to look for exoplanet transits, and then combine information from data, knowledge of the star, and from understanding of the relevant physics, to see how much we can learn about a planet so far away. Will you find a "hot Jupiter", like so many exoplanets already discovered? Or will you be one of the first to find a habitable, Earth-like planet?

Transit Method Simulation Procedure

From a current web browser, go to one of the websites given below to start running the simulation, which guides you through eight (8) pages of information and calculations.

Simulation Website

http://www.bridgewater.edu/~rbowman/ISAW/Transit-1.html

If for some reason this website is not accessible, an alternate site for the simulation is:

http://www.edtechbybowman.net/PhysAstroSims/exoplanets/Transit-1.html

Page 1: Read the Introduction on this web page – you will have to answer questions later. Then select the "Simulation" link, and you will be directed to page 1 of the exercise, which shows a patch of sky with stars, numbered 1 through 8. Pick a star to analyze. You will see a plot of the star's "light curve" (brightness as a function of time). If the star has a planet whose orbit causes transits along our line-of-sight, you will see periodic "dips" in this plot. As carefully as you can (here is where a printer will help), estimate the Period P of the planet's orbit.

Page 2: A star's "spectral type" (which is related to its mass, temperature, and power output) is given by a letter-number code. For instance, our star is a "G5", where as a lower mass star could be a "K2". (The letter sequence from high- to low- mass is OBAFGKMRN, remembered as "Oh Be A Fine Guy/Girl Kiss Me Right Now!"). From the table given, look up and enter the star's mass orbit. Select **Calculate** and the program use Kepler's 3rd law ($P^2 = a^3/M$) to calculate the size of the planet's orbit. You're now on your way to finding out more about the [simulated] exoplanet you have discovered!

Page 3: Plotting your star's mass vs. your planet's orbital distance, you can determine whether your new planet falls within the central star's Habitable Zone. This is the zone generally considered to be "not too hot, not too cold, but just right" for liquid water to exist in the presence of at atmosphere – conditions which are thought to be essential for life to develop. You may find it easier to accomplish this task by printing a paper copy of the plot provided.

Page 4 uses the central star's radius and temperature to estimate your exoplanet's temperature at its orbital distance (assuming a circular orbit). Here the temperatures are expressed in *Kelvin* above absolute zero. So, for example, $0^{\circ}C = 32^{\circ}F = 273^{\circ}K$. Planets with temperatures around $300^{\circ}K$ could be considered "habitable." Obviously, a planet's temperature will increase with the star's temperature and size, but decrease with orbital distance. Luckily for you, the computer does most of the work once you have provided the information. Look up the star's radius and temperature for its spectral type from the tables provided, enter the information and select **Calculate**.

Page 5 is an "automatic" calculation of your planet's physical size, based on the drop in light output as it transits the face of the star. Simply select **Calculate**, but be sure to read the information presented on this page.

Page 6 uses two separate "models" of a planet's density based on the planets in our solar system, to estimate the possible mass of the new exoplanet. (Unlike the "stellar wobble" methods of detection, the transit method does not give a planet's mass directly). One model uses your exoplanet's radius, the other uses its distance from the star – both of which have already been calculated. From each of the curves presented for each model, estimate your planet's density, and select **Calculate**. Do not be concerned if the two mass estimates are vastly different! With more exoplanets being found by various methods, we hope soon we can construct better models of solar system contents, instead of just relying on our own (possibly atypical) solar system.

Page 7 concludes the simulation with an "automatic" calculation of the probability of detecting such a planet. Since planetary systems will only orbit in the plane containing our line-of-sight by chance, the *Kepler* mission will only detect a fraction of all exoplanets. But by observing many stars, we should expect success in a significant number of cases.

Page 8 presents a summary of all the parameters you have calculated, which you can use to complete the summary table on the worksheet. **NOW REPEAT THE SIMULATION WITH A DIFFERENT STAR.** Note that some of the stars do not have detectable planets in this exercise, so if this is the case, make a note of the star's ID# and select another. Happy planet hunting!

Summary Table for the Two Simulated Exoplanets

Exoplanet System Information	1st Simulation	2nd Simulation
Star ID number(# 1-8) & special type (e.g. A0, G5, K2)		
Star's mass *MStar* (solar masses)		
Star's radius *RStar* (solar radii)		
Star's surface temperature *TStar* (Kelvin)		
Planet's measured orbital period *P* (yr)		
Planet's orbital size (semi-major axis) *a* (A.U.)		
Planet orbits in the habitable zone? (Yes or No)		
Planet's estimated surface temperature *TP* (K)		
Planet's radius *RP* (Earth radii)		
Planet's mass range (in Earth masses, min. to max.) e.g. "1.3 – 2.8"		
Probability of discovery (%)		

Notes
1. Some stars in this exercise do not have exoplanets that produce a good light curve with a measurable period between transits. If you select one of these, make a note of its ID number below the table, then pick another star.

2. Boxes containing the words "automatic" are not for data entry. After you have filled the other boxes with the necessary data, select **Calculate** and the computer will update these boxes and its ongoing summary of results.

3. In order to answer the questions properly, you will need to READ the background information on each page of the simulation. Much of this information is given below the **Next** button, so scroll down to read it.

4. The simulation keeps track of any calculations made, so if you need a piece of information about the star, or the exoplanet, look for it in the ongoing summary at the top of the web page. You can also record your results in the Summary Table above as you progress through the simulation or copy the numbers from the final page of the exercise.

22. Speed of Sound Discovery Lab

Introduction
Resonance occurs whenever something is forced to vibrate (by an outside object) at its own natural frequency. When this happens, the amplitude of the vibration increases dramatically. If the vibration is producing sound, we hear the increase in amplitude as an increase the volume of the sound.

The most common place to see resonance is in tubes. Most musical instruments are just tubes with two (2) open ends (open-open) or one open and one closed end (open-closed). The air in the instrument is forced to vibrate by the oscillation at the end (the reed, or the player's lips, etc.). If one of the frequencies of the vibration matches the natural frequency of the air in the tube, the air resonates and vibrates with a large amplitude, making a loud sound. (Any of the vibrations that do not match the natural frequency make the air vibrate, but with a small amplitude so you cannot hear them.)

Purpose
To change the tone (or pitch) an instrument plays the length of the tube is changed, which changes the air's natural frequency and to experiment with different tubes and change their lengths until their natural frequency matches the frequency of the tuning fork that is forcing them to vibrate

Equipment
Tuning fork, rubber mallet, and resonance tubes(i.e. two pieces of PVC pipe and one piece of PVC pipe), large graduated cylinders

Procedure
1. Measure the temperature in the laboratory by checking the thermometer in the beaker full of water on the front table:

Temperature:_____°C

2. When you come to a new station, *identify whether it is an open-open tube or an open-closed tube* and follow the correct instructions for that type of tube.

Instructions for an Open-Open Tube: There are 3 open-open tubes you will test each once and record the data in the table below.

Open-Open Tube Data

Frequency of Tuning Fork	Measured Length of Tube at Maximum Amplitude

1st Find the frequency printed on the side of the tuning fork at the station and record it in your table.

2nd Strike the tuning fork with the mallet. **FAILURE TO USE THE MALLET MAY DAMAGE THE TUNING FORK!!**

3rd Hold the vibrating tuning fork at an open end of the resonance tube so the tuning fork's vibrations are directed along the length of the tube. Prongs lined up along length of tube, as in the diagram.

4th Adjust the tube's length until you hear the volume of the sound increase to its highest amplitude. **THIS IS RESONANCE! Do not be fooled by small peaks in volume, wait until you hear a very clear increase in volume!**

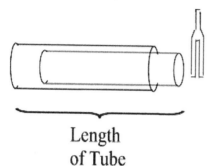

Length
of Tube

5th Measure the length of the column for which you get resonance. For an open-open tube, that is the length from one open end to the other (the combined length of both tubes). Be sure to convert your measurement into meters and record the length in your Open-Open Tube data table.

6th Take data for all of the different stations you are going to do and then wait to do the data analysis.

Instructions for an Open-Closed Tube: There are three open closed tubes record the data from in one in the table below.

Open-Closed Tube Data

Frequency Of Tuning Fork	Measured Length of Tube at Maximum Amplitude

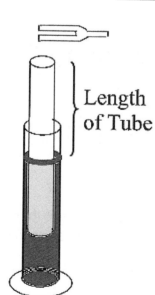

Length
of Tube

1st Find the frequency printed on the side of the tuning fork at the station and record it in your table.

2nd Strike the tuning fork with the mallet. **FAILURE TO FOLLOW THIS MAY DAMAGE THE TUNING FORK!**

3rd Hold the vibrating tuning fork at an open end of the resonance tube so the tuning fork's vibrations are directed along the length of the tube. Prongs lined up along length of tube, like in the diagram.

4th Adjust the tube's length until you hear the volume of the sound increase to its highest level. **THIS IS RESONANCE! Do not be fooled by small resonances** or peaks in volume, wait until you hear a very clear increase in volume!

5th Measure the length of the column for which you get resonance.

For an open-closed tube, measure the length from the open end to the surface of the water (this represents the closed end). Be sure to convert your measurement into meters and record the length in your Open-Closed Tube data table

6th Take data for all of the different stations you are going to do and then wait to do the data analysis.

Graph
Create a graph that compares length of the two types of tubes when resonating to the same note.

Questions
Answer the following questions by looking at your results in your data tables and your knowledge of resonance.
1. What is the *relationship* between the length of an open-open tube and the frequency of the sound that resonates in it?

2. What is the *relationship* between the length of an open-closed tube and the frequency of the sound that resonates in it?

3. If you want to have your didgeridoo, flute, trombone, etc... play a lower note, what do you have to do to the tube? Think about how a musician changes the note they are playing. If they want the note to be low in frequency what do they do compared to if they want to play a high frequency note.

4. Compare an open-open tube and an open-closed tube that resonate at close to the same frequency. How do the lengths of the tubes compare? Give the exact amount one tube will be longer than the other tube.

5. Compare an open-open tube and an open-closed tube that are approximately the same length. How do the frequencies that resonate in each tube compare? Again give the exact about greater one frequency will be than the other.

6. If you were building a pipe organ in a small church and wanted to get the lowest sound possible out of your pipes, what type of tube would you use: open-open or open-closed?

Analysis
When any object vibrates at its natural frequency or is made to resonate, there is a standing wave inside the object. In a tube full of air, that wave will be a standing sound wave. Because standing waves have to have either nodes or antinodes at the edges of the object, the length of a resonating tube can tell us the wavelength of

the sound waves. The relationship between wavelength and length of the resonating tube depends on which type of tube it is: open-open or open-closed.

Open - Open   **Open - Closed**

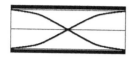 $w = 2L$  $w = 4L$

Wave Speed Equation

$s = \lambda f$ If you know the wavelength of the sound wave then all you need to know is the frequency of the sound wave and it is simple to calculate the speed of the sound wave using the wave-speed equation. We will use our knowledge of resonance and standing waves to find the speed of sound several times and see how accurately we can get. With a little bit of effort, you can get very accurate values for the speed of sound in air so try to make your measurements as accurate as you can.

Compare the data to the expected value of the speed of sound.

Expected Speed of Sound

Previous research gave a formula to calculate the expected speed of sound:
s = 332 m/s + 0.6 m/s (**T**), where s is the speed of sound and T is the temperature in degrees Celsius.

s = _____

Calculations

Calculated Wavelength for the Note	Calculated Speed of Sound	Percent of Error

1. For each tube and tuning fork you measured, use the appropriate tube equation to find the wavelength based on the measured length of the resonating tube. Use the open-open tube equation for open-open tubes and the open-closed tube equation for open-closed tubes. Record these calculations in the Calculations tables.

Calculated Wavelength for the Note	Calculated Speed of Sound	Percent of Error

2. Once you have the frequency and wavelength, use the wave speed equation to

calculate the speed of the sound wave based on its wavelength and frequency.

3. After you have calculated the speed of sound for all of your trials, average your results for each type of tube.

4. Calculate the percent of error for the average speed on each type of tube.

Questions

Answer the following questions using the results in your data table and the equations for both types of tubes.

1. Which type of tube gave you the lowest percent error, open-open or open-closed? Why do you think there was a difference?

2. How long would a trombone (what type of tube is a trombone?) have to be to produce a 3.4 m long sound wave?

3. How long would the sound-making portion of a flute be when you make a sound wave with a frequency of 900 waves/sec? (Assume speed of sound = 343 m/s)

4. What frequency sound wave will you get from a jug that is .4 m (40 cm) tall? (Assume speed of sound = 343 m/s)

Appendix A   Frequently Used Constants

Proton mass, $m_p = 1.67x10^{-27}kg$

Neutron mass, $m_n = 1.67x10^{-27}kg$

Electron mass, $m_e = 9.11x10^{-31}kg$

Speed of light, $c = 3.00 \ x \ 10^8 \ \frac{m}{s}$

Electron charge magnitude, $e = 1.60x10^{-19}C$

Coulomb's law constant, $k = \frac{1}{4\pi\varepsilon_0} = 9.0x10^9 N \cdot \frac{m^2}{C^2}$

Universal gravitational constant, $G = 6.67 \ x \ 10^{-11} \ \frac{m^3}{kg \cdot s^2}$

Acceleration due to gravity at Earth's surface, $g = 9.8 \ \frac{m}{s^2}$

Physics Lab Manual

Appendix B   Frequently Used Formulas

Mechanics

$v_x = v_{x0} + a_x t$

$x = x_0 + v_{x0}t + \frac{1}{2}a_x t^2$

$v_x^2 = v_{x0}^2 + 2a_x(x - x_0)$

$\vec{a} = \frac{\Sigma \vec{F}}{m} = \frac{\overrightarrow{F_{net}}}{m}$

$|\overrightarrow{F_f}| \leq \mu |\overrightarrow{F_n}|$

$a_c = \frac{v^2}{r}$

$\vec{p} = m\vec{v}$

$\Delta \vec{p} = \vec{F}\Delta t$

$K = \frac{1}{2}mv^2$

$\Delta E = W = F_\parallel d = Fd\cos\theta$

$P = \frac{\Delta E}{\Delta t}$

$\theta = \theta_0 + w_0 t + \frac{1}{2}at^2$

$\omega = \omega_0 + at$

$x = A\cos(2\pi ft)$

$\vec{a} = \frac{\Sigma\vec{\tau}}{I} = \frac{\vec{\tau}}{I}$

$\tau = r_\perp F = rF\sin\theta$

$L = I\omega$

$\Delta L \geq \tau\Delta t$

$K = \frac{1}{2}I\omega^2$

$|\overrightarrow{F_s}| = k|\vec{x}|$

$U_s = \frac{1}{2}kx^2$

$\rho = \frac{m}{V}$

$\Delta U_g = mg\Delta y$

a = acceleration
A = amplitude
d = distance
E = energy
f = frequency
F = force
I = rotational inertia
K = kinetic energy
k = spring constant
L = angular momentum
ℓ = length
m = mass
P = power
p = momentum
r = radius or separation
T = period
t = time
U = potential energy
V = volume
v = speed
W = work done on a system
x = position
y = height
α = angular acceleration
μ = coefficient of friction
θ = angle
ρ = density
τ = torque
ω = angular speed

$T = \frac{2\pi}{\omega} = \frac{1}{f}$

$T_s = 2\pi\sqrt{\frac{m}{k}}$

$T_p = 2\pi\sqrt{\frac{\ell}{g}}$

$|\vec{F_g}| = G\frac{m_1 m_2}{r^2}$

$\vec{g} = \frac{\vec{F_g}}{m}$

$U_G = -\frac{Gm_1 m_2}{r}$

Electricity

$|\vec{F}_E| = k\left|\frac{q_1 q_2}{r^2}\right|$

$I = \frac{\Delta q}{\Delta t}$

$R = \frac{\rho\ell}{A}$

$I = \frac{\Delta V}{R}$

$P = I\Delta V$

$R_s = \sum_i R_i$

$\frac{1}{R_p} = \sum_i \frac{1}{R_i}$

Waves

$\lambda = \frac{v}{f}$

A = area
F = force
I = current
ℓ = length
P = power
q = charge
R = resistance
r = separation
t = time
V = electric potential
ρ = resistivity

f = frequency
v = speed
λ = wavelength

Appendix C   Frequently Used Units

Unit Symbols			Prefixes		
meter	m		10^{12}	tera	T
kilogram	kg		10^{9}	giga	G
second	s		10^{6}	mega	M
ampere	A		10^{3}	kilo	k
kelvin	K		10^{-2}	centi	c
hertz	Hz		10^{-3}	milli	m
newton	N		10^{-6}	micro	μ
joule	J		10^{-9}	nano	n
watt	W		10^{-12}	pico	p
coulomb	C				
volt	V				
ohm	Ω				
degree Celsius	°C				

Appendix D Greek Alphabet (Upper and Lower Case)

Upper & Lower Case	English Transliteration	Name of Letter
Aα	a	alpha
Bβ	b	bēta
Γγ	g	gamma
Δδ	d	delta
Eε	e	epsilon
Zζ	z	zēta
Hη	ē	ēta
Θθ	th	thēta
Iι	i	iota
Kκ	k	kappa
Λλ	l	lamda
Mμ	m	mu
Nν	n	nu
Ξξ	x	xi
Oο	o	omicron
Ππ	p	pi
Pρ	r (rh when initial)	rho
Σς	s	sigma
Ττ	t	tau
Υυ	u	upsilon
Φφ	phi	phi
Χχ	ch	chi
Ψψ	ps	psi
Ωω	ō	ōmega

Index

Made in the USA
Coppell, TX
30 August 2021